*Oliver Geschke, Henning Klank,*
*Pieter Telleman*
**Microsystem Engineering**
**of Lab-on-a-chip Devices**

*Oliver Geschke, Henning Klank,*
*Pieter Telleman*

# Microsystem Engineering
# of Lab-on-a-chip Devices

**WILEY-**
**VCH**

WILEY-VCH Verlag GmbH & Co. KGaA

Editors

**Dr. Oliver Geschke**
**PhD Henning Klank**
**Prof. Pieter Telleman**
Micro- and Nanotechnology Center (MIC)
at the Technical University of Denmark
DTU Building 345 east
Ørsteds Plads
DK-2800 Kgs. Lyngby
Denmark
www.mic.dtu.dk


Contributors

**PhD Henrik Bruus**
**Goran Goranovic**
**PhD Anders Michael Jorgensen**
**Dr. Jörg P. Kutter**
**PhD Klaus Bo Mogensen**
**Gerardo Perozziello**
**Daria Petersen**
all:
Micro- and Nanotechnology Center (MIC)
at the Technical University of Denmark
DTU Building 344 east
Ørsteds Plads
DK-2800 Kgs. Lyngby
Denmark
www.mic.dtu.dk

**Library of Congress Card No.: applied for**

**British Library Cataloguing-in-Publication Data**
A catalogue record for this book is available from
the British Library.

**Bibliographic information published
by Die Deutsche Bibliothek**
Die Deutsche Bibliothek lists this publication
in the Deutsche Nationalbibliografie; detailed
bibliographic data is available in the Internet at
<http://dnb.ddb.de>

© 2004 WILEY-VCH Verlag GmbH & Co. KGaA,
Weinheim

Printed in the Federal Republic of Germany
Printed on acid-free paper

**Typesetting**   K+V Fotosatz GmbH, Beerfelden
**Printing**   Strauss Offsetdruck GmbH, Mörlenbach
**Bookbinding**   Litges & Dopf Buchbinderei GmbH,
Heppenheim

**ISBN**   3-527-30733-8

# Contents

# Preface

We live in a world that is influenced by technological developments. One of the clearest examples of this is microtechnology. The use of microtechnology to miniaturize and functionally integrate electronic components has changed our world and hardly any facet of our lives is not in some way affected by microelectronics. Building on the experience of microelectronics research and industry we have started to apply microtechnology to chemistry and biochemistry. We stand to gain many advantages including improved performance, portability, and reduction of cost. The application of microtechnology to chemical and biochemical analysis is a very multidisciplinary topic which needs input from scientist and engineers with different backgrounds. This book combines the experience of a group of engineers, chemists, physicists, and biochemists who are applying microtechnology to chemical and biochemical analysis at the Mikroelektronik Centret (MIC) at the Technical University of Denmark (DTU). The various stages in the development of such microsystems are described in this text book: from concept to design, to fabrication, and to testing. There is little doubt in the international research and industry community that the application of microtechnology to chemistry and biochemistry will revolutionize our lives in a way that is comparable to what we have seen with microelectronics. Our aim with this book is to allow a broad range of scientists and engineers to get interested and familiarized with this very exciting topic.

Lyngby, July 2003

*Oliver Geschke*
*Henning Klank*
*Pieter Telleman*

# 1
# Introduction
PIETER TELLEMAN

## 1.1
## Learning from the Experiences of Microelectronics

Try to think back to the time that your parents were your age and imagine the technological developments that have taken place since then. Sometimes it is hard to
imagine that only 2 decades ago personal computers, mobile phones, compact disks
(CD) players, and digital video disks (DVD) players did not exist. What made these
technological developments possible? One of the major contributing factors is microelectronics. The first breakthrough from electronics to microelectronics was the invention of the transistor in 1947 at Bell laboratories. Transistors provided a better,
cheaper alternative to mechanical relays, which were the standard electronic component for switching and modulating electronic signals. With improving semiconductor technology, transistors became progressively smaller, cheaper, and better. A second breakthrough was the introduction of the integrated circuit in 1959, by which
numerous transistors and other electronic components together with the necessary
wiring were organized on a thin silicon disk or wafer. In 1965, only 4 years after the
introduction of the integrated circuit, Gordon Moore predicted an exponential
growth of the number of transistors in an integrated circuit (Moore's Law). Although
the pace has slowed down a bit in recent years, experts agree that the current rate of a
doubling every 18 months will continue at least for 2 more decades. If we should
summarize the process that made microelectronics so successful, we could say that
it was the combination of miniaturization, i.e., microfabrication of transistors and
other electronic components, and functional integration, i.e., the organization of
many different miniature electronic components to form integrated circuits with
complex functions. Since the application of miniaturization and functional integration to electronics, the same strategy has been applied to a range of other disciplines,
e.g., mechanics and optics. One example of a microelectromechanical system
(MEMS) is the accelerometer. The deployment of airbags in cars depends on signals
from a number of accelerometers, i.e., miniaturized mechanical sensors that measure the g forces on the car. Other examples of MEMS are pressure sensors and mi-
crophones. The promise of faster and better data transfer offered by optical communication has resulted in the application of microtechnology to develop microstructures for the manipulation of light, e.g., micromirrors and optical switches.

*Microsystem Engineering of Lab-on-a-chip Devices*
O. Geschke, H. Klank, P. Telleman
Copyright © 2004 Wiley-VCH Verlag GmbH & Co. KGaA, Weinheim
ISBN: 3-527-30733-8

In 1979, S.C. Terry et al. presented "A gas chromatographic air analyzer fabricated on silicon wafer using integrated circuit technology" [1]. This was the first publication that discussed the use of techniques borrowed from microelectronics to fabricate a structure for chemical analysis . The introduction of the concept of micro total-analysis systems (μTAS) by Manz and coworkers in 1990 [2] triggered rapidly growing interest in the development of microsystems in which all the stages of chemical analysis such as sample pre-preparation, chemical reactions, analyte separation, analyte purification, analyte detection, and data analysis are performed in an integrated and automated fashion. The aim of this textbook is to provide you with a comprehensive understanding of the concept of μTAS. We will introduce you to microfluidics, i.e., the manipulation of small amounts of reagents and sample on microchip, simulation and modeling of microfluidics, fabrication of microsystems for chemical analysis in silicon, glass, and plastics, packaging of microsystems, and several examples of chemical analysis in microstructures.

## 1.2
## The Advantages of Miniaturizing Systems for Chemical Analysis

Why is it that, when the concept of μTAS was introduced in the early 1990s, it attracted so much interest from the scientific and the industrial community? It was because the conventional approach to chemical analysis can no longer meet all the requirements that many applications demand. Let us look at some of these requirements and see how μTAS can offer unique solutions.

With rapid developments and growing interest in, e.g., medicine, drug discovery, biotechnology, and environmental monitoring, we have become more and more dependent on chemical analysis. Traditionally, chemical analyses have been performed in central laboratories because they require skilled personnel and specialized equipment. However, the trend is to move chemical analysis closer to the 'customer'. Some examples are pregnancy tests, blood glucose concentration tests for diabetes patients, and analysis of soil and water samples. These chemical test kits can be acquired off the shelf and can be used in the home by persons with no special training in chemistry. This trend of decentralization of chemical analyses is expected to continue. For this to happen we need to make analytical equipment smaller and thus portable, easier to operate, and reliable. The results of the chemical analyses must be processed so that it is easy for the user to interpret. The concept of μTAS builds on performing all the necessary steps that are required for a chemical analysis on a miniaturized format and thereby offers portability. Because the microfabricated components in a μTAS can be operated with very low power consumption, battery-operated analytical equipment opens up the possibility of performing chemical analyses in the field independent of a power grid. Automation of the entire chemical analysis process and data processing is also part of the μTAS concept. In its extreme case μTAS can be represented as a black box where the user needs only to apply the sample and push a start button to perform the chemical analysis and retrieve the results. Microfabrication allows us to reproduce the same carefully designed µTAS many times with the same specifications. When care is taken to address reliability at the stage of designing a µTAS, reliability can be warranted for large batches. At the heart of each µTAS is a chip in which fractions of microliters of samples and reagents are moved around with very high accuracy. Traditionally chemical analyses are performed by mixing milliliters of samples and reagents in conventional test tubes and analyzing the product in an analytical instrument, e.g., a spectrophotometer. Especially when the samples and reagents are in short supply or very expensive, µTAS offers a significant decrease in costs by dramatically reducing the volume of samples and reagents that are needed to perform a chemical analysis. We already mentioned that once a µTAS has been successfully developed, it can be reproduced faithfully in very large numbers. This opens up the possibility of processing samples in parallel, which is very useful when the same chemical analyses must be performed many times over. This is exactly what drug discovery is about. A drug candidate often needs to be identified from a pool of many thousands of samples by performing a particular chemical analysis on each sample (this process is referred to as high-throughput screening or HTS). Today HTS is implemented by performing the chemical analysis in microtiter plates in combination with robotic handling of the samples and reagents. The possibility µTAS offers of parallelizing chemical analyses is seen as an interesting alternative to the use of microtiter plates and will eventually allow an increase in throughput.

Often, we want to know how the concentration of an analyte changes in time, i.e., online monitoring. It is better to continuously monitor the concentration of glucose in the blood of a diabetes patient than to measure the glucose concentration once every so many hours. Continuous analysis of ammonium in wastewater is more valuable for controlling a sewage-treatment plant than a measurement only 2 or 3 times a day. With conventional methods of chemical analysis is it difficult to implement online chemical analyses. Handling and processing of the sample is, at least in part, done manually and often in specialized laboratories. But with µTAS, we can bring the chemical analyses close to the place where they need to be performed, independent of a laboratory and laboratory personnel. Sample handling and processing, the chemical analysis, and data processing are integrated in µTAS, which makes it very well suited for online measurements.

The advantages of µTAS can be summarized as follows: µTAS offers portability, reliability, reduction of sample and reagent consumption, automation of chemical analysis, high-throughput screening, and online analysis. Keep in mind however, that µTAS has been around only since the late 1980s and that a much research and development still has to be performed in order to fully benefit from all its advantages. Several issues that are essential to the widespread use of µTAS have received little attention so far. The most prominent of these issues are interconnection and packaging. Regardless of how skilled we are in designing and fabricating µTAS, the chip at the heart of the µTAS must be interfaced to the macroworld of the user. For µTAS, this requires fluidic, mechanic, optical, and electronic interconnections. Furthermore, µTAS must be packaged so they can be handled safely without damaging the delicate microstructures on the chip. Both issues must be dealt with to allow for successful commercialization and thereby wider use of the technology.

## 1.3
## From Concept to μTAS

When you received this book you most likely started to flip through the pages to see what you can expect in the coming days or weeks. And you discovered that this book addresses a wide range of subjects that belong to many different disciplines, including physics, chemistry, and computer sciences. μTAS is a truly multidisciplinary activity that requires input from scientists having many different backgrounds.

The process of developing a μTAS consists of several discrete steps, starting with determining the specifications for the μTAS (Fig. 1.1). These specifications depend mainly on the nature of the chemical analysis and must answer questions such as: which reagents are used? what are the reaction kinetics? at what temperature are the reactions performed? what means of detection will be used? what is the desired range of detection? what is the required limit of detection? The chemistry in turn determines what material can be used for fabrication of the μTAS, for example: should it be transparent? are the reagents aggressive? is the μTAS intended for single use or multiple use? Inherent to combining mechanics, fluidics, optics, and electronics in μTAS is the formation of interfaces between these media. One must be aware of the fact that the sensor function of μTAS is actually based on the interfaces between 2 or more media, e.g., for absorption measurements you need an interface between light and a chemical. The interface of μTAS and the user, i.e., interconnection and packaging, must be also considered during

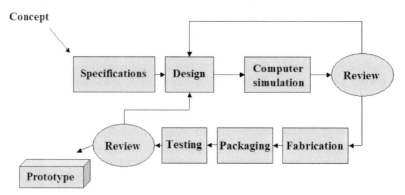

**Fig. 1.1** From concept to μTAS. The successful development of a μTAS involves a number of discrete steps: specifications of the chemical analysis, design, modeling to evaluate performance, fabrication, and testing. Reviews of the modeling and test results enable optimization of the performance of the μTAS.

the specification phase. Defining the specifications for μTAS is a process that should involve all project members because it affects the overall μTAS performance.

With the specifications in place, the next step is to design the μTAS. Design constitutes the most important block in the flow sheet from μTAS concept to prototype and is discussed in more detail in chapters 3 and 4. It is here that considerations of μTAS concept, definition of interfaces, and specifications are translated to a fabrication plan. Developing a sequence of process steps for individual μTAS components, e.g., micropumps, is challenging in itself, but aiming at μTAS, where the entire process sequence involves a variety of integrated components, raises questions of process sequence and compatibility. How does one combine, from a process point of view, for example, microfluidic components with optical components without loosing the properties of the individual structures due to process incompatibility somewhere along the way? Is the choice of a particular process sequence compatible with demands for packaging? One of the first steps in establishing a complete and effective μTAS platform must be the categorizing of all process steps that are involved in making individual components, investigating process compatibility, and finding alternative processes or process sequences in cases of incompatibility. Design is in many ways a matter of experience and intuition and, with a design that satisfies the demands of the different partners involved, it is in principle possible to start fabricating the μTAS. However, depending on the complexity of the design, it is often very difficult to predict the performance of the μTAS intuitively. In these cases computer simulations may provide a means to study the performance of a μTAS prior to fabrication.

Computer simulations can significantly shorten the possibly long process of μTAS design, fabrication, and testing. The behavior of individual components, as well as the interplay between integrated components, can be predicted by computer simulations. By including a review step after computer simulation, structures can be optimized for their geometry and operational parameters based on the simulation results prior to actually fabricating the components or devices. This rational approach constitutes a significant improvement over the approach in which computer simulation is omitted and structures are optimized by numerous rounds of fabrication and testing. Important aspects of computer simulations are addressed in chapter 5. Key to the development of μTAS is microfabrication: the fabrication of structures down to micrometers in size. Aspects of microfabrication in silicon, glass, and polymers are discussed in chapters 6, 7 and 8. The explosive growth of microelectronics has led to a wide range of microfabrication tools for silicon, and consequently, much higher levels of experience and expertise exist for working with silicon as a material for microtechnology. Silicon presented an obvious choice as a material for the microelectronics industry due to its semiconductor properties. Few materials can surpass silicon when it comes to fabricating microstructures: silicon is suitable for the fabrication of electronic, mechanical, and optical components and thereby allows for high levels of functional integration. However, the superiority of silicon as a material for μTAS is debatable because the chemical stability of silicon is not very good. In fact, many of the microfabrication methods available today are based on the controlled removal of silicon by chemical treatments. Although the surface of silicon can be treated to withstand harsh chemical environments, other materials may be more suitable for certain applications. Another important argument for investigating alternative materials is the relatively high cost of silicon, especially in applications where µTAS that have been in contact with biohazardous materials like blood are discarded after a single use. For these reasons polymers and glasses offer interesting alternatives to the use of silicon for µTAS. Because the use of polymers and glasses for mechanical, optical, and electronic components is still very much under development, fabrication of these materials carries with it concessions as to the level of functional integration that can be achieved. Hybrid solutions, in which microstructures of different functions and fabricated of different materials are assembled to make up a complete µTAS, will most likely arise.

With fabrication complete, structures must be tested in the laboratory to assess to what extent they live up to the previously defined specifications and how well computer simulations were able to predict the performance of the µTAS. When the device does not perform according to the specifications, all aspects downstream from the specifications need to be reconsidered. Modeling tools will have to be modified if they cannot predict the behavior of µTAS accurately enough.

As mentioned earlier, the aim of µTAS is a complete integration of all necessary steps for conducting a complete chemical analysis. Depending on the duration and complexity of the entire process of design and fabrication of µTAS, you can imagine that the final µTAS can be very expensive. In applications where the µTAS offers a significant improvement over conventional chemical analysis techniques and where the expected useful lifetime of the µTAS is long, the potential high cost of µTAS may not be the decisive factor that prevents its use. However, in applications where the µTAS is discarded after a single use, the cost of µTAS is very important. In some cases we may be simply unable to realize a true µTAS because we lack the technology to integrate certain essential components, e.g., lasers. The formal concept of functional integration in µTAS and all the accompanying advantages must therefore be balanced against complexity, cost, and feasibility. Undoubtedly we will see many examples of µTAS that result from the assembly of a microfabricated chip with conventional, possibly miniaturized, components, e.g., pumps, light sources, electronics. The assembly of these hybrids between microtechnology and conventional technology can be adjusted so that the level of integration makes sense for the individual application. With hybrid technology, you can discard certain parts of the hybrid while keeping expensive functional units like pumps and light sources.

At the time of writing this textbook, the commercial market for µTAS-based products is still rather small. However, market research reports predict consistent growth in the global market for µTAS-based products. These reports also agree that chemistry and the life sciences continue to be the major users of microsystem technology. With the anticipated future technological developments in chemistry and the life sciences, it is clear that microtechnology in general and µTAS specifically will play an essential role in these developments. Many fundamental problems still need to be addressed to allow for the routine application of µTAS in chemistry and the life sciences, the most pertinent being interconnection and packaging of the µTAS to allow handling by the operators. It is likely that answers to these 2 factors will determine the ultimate commercial success of µTAS. Interconnection and packaging are discussed in detail in chapter 9. The need for a paradigm shift in chemical and biochemical analyses to satisfy the needs of research and industry is, however, so large that solutions to these problems will undoubtedly be found and µTAS will be a part of our future.

## 1.4
## References

1 TERRY, S.C., JERMAN, G.H., ANGELL, J.B. (1979) *A gas chromatographic air analyzer fabricated on a silicon wafer.* IEEE Trans. Electron Devices. ED-26, 12, 1880–1886.

2 MANZ, A., GRABER, N., WIDMER, H.M. (1990) Miniaturized total chemical analyses systems. A novel concept for chemical sensing. *Sens. Actuators, B Chem.*, B1 (1–6), 244.

# 2
# Clean Rooms
DARIA PETERSEN and PIETER TELLEMAN

The functional components in μTAS can have dimensions down to micrometers, and particles in the atmosphere that contaminate such components can completely destroy the function of a μTAS. The concentration of particles larger than 0.5 μm in the air of a classroom or office building can be as high as 50 million particles per cubic meter. To avoid contamination of wafers with particles, a special laboratory is needed: a clean room. In a clean room, air-borne particles are removed by continuously filtering the air through a high-efficiency particulate air filtering (HEPA) system. HEPA filtering systems are a class of air filters that retain close to 100% of particles as small as 0.3 μm (Fig. 2.1).

Clean rooms include several sections that have different requirements for cleanliness. The air in areas where wafers are handled must be kept as clean as possible, but the air quality in service areas, which contain the bulk of the equipment, is less critical. To prevent contaminated air from entering those areas where air quality is most critical, the air pressure in these areas is kept slightly higher. To minimize the circulation of particles in the clean room, filtered air enters the clean room through the perforated ceiling and is removed through the raised, perforated floor. The air flow regime in a clean room is laminar, i.e., turbulent flow is absent, to prevent particles from travelling through the clean room so they can be efficiently and rapidly removed. The velocity of the laminar flow in the clean room that is used at the Mikroelektronik Centret at the Technical University of Denmark is about 0.4 m s$^{-1}$, the intake of fresh air into this clean room is about 30 000 m$^3$ h$^{-1}$, and the air flow inside this clean room is about 130 000 m$^3$ h$^{-1}$. Air inside the clean room is recycled as much as possible, but exhaust air from the equipment, fume hoods, and wet chemical benches is not recycled. To avoid contamination, equipment is placed in clean rooms so as to prevent return air paths. Only the purest starting materials and processing chemicals should be used in a clean room, and equipment in the clean room must be cleaned periodically to maintain the cleanest environment possible. Clean room users are responsible for daily cleaning and releasing as few particles as possible. For example, one has to avoid quick movements in the clean room so as not to disturb the laminar air flow pattern. Talking to colleagues in the clean room must be kept to a minimum, because talking generates many particles and aerosols. Smoking within half an hour before entering a clean room is forbidden, to reduce the emission of

*Microsystem Engineering of Lab-on-a-chip Devices*
O. Geschke, H. Klank, P. Telleman
Copyright © 2004 Wiley-VCH Verlag GmbH & Co. KGaA, Weinheim
ISBN: 3-527-30733-8

**Fig. 2.1** Ventilation system of the clean room at the Mikroelektronik Centret at the Technical University of Denmark,
which contains HEPA filters to remove particles from the air.

small smoke particles from the lungs. No form of makeup is allowed while working in the clean room, since most makeup is based on particles. Humidity and
temperature in clean rooms are kept constant at 45% relative humidity and 21 °C,
to maintain consistent experimental conditions and create a good working environment.

Federal Standard 209E of the USA describes basic design and performance requirements for different classes of clean room. This standard is used for most
clean rooms. Classification of a clean room by this federal standard sets the maximum number of particles larger than 0.5 μm in each cubic foot of air. For example, a class-1000 clean room has fewer than 1000 particles larger than 0.5 μm per
cubic foot. For the microfabrication of μTAS, class-1000 or class-100 clean rooms
are usually sufficient.

People working in the clean room are the main source of contamination. A person can shed as many as tens of thousands to tens of millions of particles per

**Fig. 2.2** A cleanroom suit prevents contamination of the clean room by the user.

minute. To prevent the bulk of these particles from entering the clean room, individuals working in the clean room have to put on a cleanroom suit before they enter the clean room. A cleanroom suit consists of a cap, a coverall, and boots. In addition to the cleanroom suit, disposable gloves are used, which means that the suit covers pretty much the entire body except for the face (Fig. 2.2).

Cleanroom suits are cleaned frequently and made from materials that emit very few particles. Microfabrication involves extensive use of many toxic gases and other dangerous chemicals and therefore many safety precautions are in force in a clean room. Keep in mind that the cleanroom suit is there only to protect the clean room from the user and not the other way round. Cleanroom suits offer no protection from chemicals. When handling chemicals, additional protection is offered by face shields and special chemical-resistant gloves.

A clean room is a complex and potentially dangerous laboratory environment, which requires that everyone who will work there must be trained extensively in proper cleanroom behavior and proper safety routines.

# 3
# Microfluidics – Theoretical Aspects

Jörg P. Kutter and Henning Klank

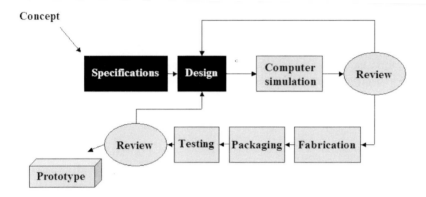

When we think about flows in everyday life and our typical experiences with them, we think of a river flowing down its bed, water flowing out of a faucet, or a beverage filling a glass. We might even think of blood flowing through our veins and maybe ink flowing out of a pen (especially when more ink is coming out than is supposed to). It is much less likely, however, that we immediately associate flow with ketchup as it oozes out of the bottle at a painstakingly slow pace, only to suddenly splash out in large quantities after the irritated diner customer administers a slap to the bottom of the upturned bottle. And yet, this also is flow behavior, and actually normal behavior for a liquid such as ketchup.

From these everyday experiences, where we look at flows on the centimeter, meter, and perhaps kilometer scale (lakes, oceans) we would not immediately expect liquids to behave any differently if we observed them on smaller scales – at the millimeter or even micrometer level. And yet that is exactly what happens! In this chapter we will see how many phenomena that we are so used to living with, and which we take for granted, have next to no significance for fluids in the micro-world: inertia means nothing on these small scales, but viscosity rears its (hideous) head and becomes a very important player. The (seemingly) random and chaotic behavior of flows in our experience is reduced to much more well-behaved and 'smooth' (laminar) flows in the smaller domains. And diffusion, on larger

*Microsystem Engineering of Lab-on-a-chip Devices*
O. Geschke, H. Klank, P. Telleman
Copyright © 2004 Wiley-VCH Verlag GmbH & Co. KGaA, Weinheim
ISBN: 3-527-30733-8

scales an almost ridiculously ineffective transport mechanism, suddenly becomes the dominant process, a strong ally or a mighty opponent, mostly depending on what you want to achieve. And, finally, surfaces become an ever more important factor to reckon with. The ratio of surface to volume increases drastically as dimensions are reduced, going from a value of $0.006 \, \text{m}^{-1}$ for a cube of side length $1 \, \text{km}$ to $6 \, \text{m}^{-1}$ for a cube of side length $1 \, \text{m}$, and further to $6000 \, \text{m}^{-1}$ for a cube of side length $1 \, \text{mm}$, and finally to $6\,000\,000 \, \text{m}^{-1}$ for a cube of side length $1 \, \mu\text{m}$. Again, this is a 2-sided coin, where some applications greatly benefit from an increased surface to volume ratio, while in others such phenomena as adsorption become increasingly harder to deal with.

The goal of this chapter is to give you some basic insights into flow behavior at small scales and on the most important processes and phenomena that must be kept in mind when attempting to design microsystems for chemical analyses or reactions. You are of course encouraged to consult specialized books for more in-depth information (ample references are given here), but by the time you have finished this chapter you should have acquired a good primary understanding of the issues involved in designing microfluidic systems.

## 3.1
## Fluids and Flows

Typically, a fluid can be defined as a material that deforms continually under shear stress, i.e., the application of an external force attempting to displace part of the fluid elements at a boundary layer (i.e., the surface). In other words, a fluid can flow and has no rigid three-dimensional structure. For all practical purposes, the fluids we encounter in everyday life are gases (air or its components) and liquids (water, oil, syrup, …). More complex systems consisting of several phases can also be classified as fluids (blood, suspensions, emulsions, …). Fluid behavior has been studied extensively for several centuries and a number of monographs and articles have been published on the subject (see, e.g., [1-5]). In the following, we will focus on some of the most important aspects of a fluid and its physical behavior. Since we almost exclusively deal with liquid systems in this book, the remainder of this discussion will focus on liquids only.

Three important parameters characterizing a liquid are its density, $\rho$, the pressure of the liquid, $P$, and its viscosity, $\eta$.

The density is defined as the mass, $m$, per unit volume, $V$:

$$\rho = \frac{m}{V} \tag{3.1}$$

Typical values for several fluids are listed in Tab. 3.1. We will encounter the density again in several issues, including the definition of the kinematic viscosity, discussions of surface tension and capillary forces, and when looking at the buoyancy of particles immersed or suspended in a liquid.

**Tab. 3.1**  Densities of some common fluids in (g cm$^{-3}$) [6, 7].

| Fluid | Temperature, °C | Density |
|---|---|---|
| Water | 0 | 0.999 |
| Water | 20 | 0.998 |
| Ethanol | 0 | 0.806 |
| Ethanol | 20 | 0.791 |
| Ether | 0 | 0.736 |
| Olive oil | 20 | 0.92 |
| Glycerin | 0 | 1.26 |
| Mercury | 0 | 13.60 |
| Mercury | 20 | 13.55 |
| Air | 20 | 0.0012 |

The pressure in the liquid is dependent only on the depth (i.e., the pressure increases when going from the surface to the bottom of a lake or to greater depths in an ocean), and it is the same at every point having the same elevation. Furthermore, the pressure is not affected at all by the shape of the vessel containing the liquid. In a planar microsystem with channel depths of a few micrometers to a few hundred micrometers, pressure differences because of different depths are not an issue and can be neglected. However, since these channels are not closed systems, but have inlets and outlets, any pressure difference induced externally at these openings is transmitted to every point in the liquid, thereby inducing the liquid to flow. Such external sources are, for example, differences in the filling height of liquid reservoirs connected to the channels or surface tension effects. Pumping in microsystems by making use of such phenomena is discussed in more detail in Chapter 4.

In many applications of microsystems, spherical particles, such as chemically functionalized silica or latex beads or entire living cells, are immersed in the liquid within the microchannels. Depending on the particle's density compared to the density of the liquid it is immersed in, the particle floats, sinks, or has neutral buoyancy. This is summarized in Archimedes' principle: the buoyant force acting on an immersed body is equal in magnitude (but opposite in direction) to the force of gravity on the displaced fluid. Thus, if a particle has a density greater than that of the liquid (meaning that a comparable volume is heavier) it sinks, or sediments. If its density is less than that of the liquid it floats. Fish (using their swim bladder) and submarines (using air tanks) can adjust their average density to be equal to that of seawater, thereby achieving neutral buoyancy.

These parameters and behaviors are obtained in experiments performed on stationary liquids. But if we attempt to set the liquid into motion, we experience a resistance to our effort, resembling an internal friction, also called *viscosity*. To better understand the concept of viscosity we assume the following model (Fig. 3.1): two solid parallel plates are set on top of each other with a liquid film of thickness $L$ between them. The lower plate is stationary, and the upper plate can be set into motion by a force, $F$, resulting in velocity, $v$. The movement of the upper plane first sets the immediately adjacent layer of liquid molecules into motion; this layer transmits

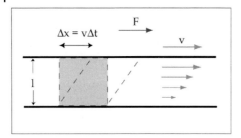

**Fig. 3.1** Schematic model explaining viscosity of fluids

the action to the subsequent layers underneath it because of the intermolecular forces between the liquid molecules. In a steady state, the velocities of these layers range from $v$ (the layer closest to the moving plate) to 0 (the layer closest to the stationary plate). The applied force acts on an area, $A$, of the liquid surface (surface force), inducing a so-called shear stress ($F/A$). The displacement of liquid at the top plate, $\Delta x$, relative to the thickness of the film is called shear strain ($\Delta x/L$), and the shear strain per unit time is called the shear rate ($v/L$). Finally, we can define the coefficient of viscosity, $\eta$, as the ratio of the shear stress to the shear rate:

$$\eta = \frac{F/A}{v/l} \tag{3.2}$$

Equivalently, we can say that $\eta$ is the proportionality factor between the shear stress and the shear rate. In the definition given in Eq. 3.2 we made the simplification that the velocity varies linearly with position. In more general terms, we would have to define $\eta$ as

$$\eta = \frac{F/A}{dv/dy} \tag{3.3}$$

Typical values of $\eta$ for several liquids are shown in Tab. 3.2.

**Tab. 3.2** Viscosities of some common fluids in $10^{-2}$ ($g \cdot cm^{-1} \cdot s^{-1}$) [6, 7].

| Fluid | Temperature, °C | Viscosity |
|---|---|---|
| Water | 0 | 1.787 |
| Water | 20 | 1.002 |
| Water | 100 | 0.282 |
| Ethanol | 0 | 1.773 |
| Ethanol | 20 | 1.200 |
| Acetone | 25 | 0.316 |
| Blood | 37 | 4.5–5.5 |
| Olive oil | 10 | 138.0 |
| Olive oil | 20 | 84.0 |
| Mercury | 20 | 1.554 |
| Air | 18 | 0.018 |

As defined in Eq. 3.3, the viscosity is independent of the applied shear stress (it is, however, temperature dependent). Such a liquid is also called a *Newtonian liquid*. Water, oil, glycerin, etc., are Newtonian liquids, meaning that the viscosity is constant, independent of the applied shear stress. In other liquids, however, the viscosity changes with the applied shear stress; these liquids are consequently called *non-Newtonian*. The viscosity can change to lower (shear-thinning) or higher (shear-thickening) values with increasing shear stress. A well known example of a non-Newtonian liquid with shear thinning is ketchup, which should be viscous at low shear rates (i.e., it should stay on your chips), but be easily moved at higher shear stress (i.e., when you scoop it up from the plate). Another important non-Newtonian liquid everyone is familiar with is blood.

With this knowledge of density and viscosity, we can now proceed to look further at liquids in motion. What kind of flow will we get in microsystems, in channels that have typical dimensions on the micrometer scale? Will we get flows similar to what we know from everyday life? Unpredictable, chaotic flows, or steadier, more well-behaved flows? To answer this question we can find the relation between the magnitudes of the inertial and viscous forces. Using the expression given in Eq. 3.4, we obtain the dimensionless Reynolds number, *Re*:

$$\text{Re} = \frac{\rho d v}{\eta} \tag{3.4a}$$

where $d$ is the typical length scale (e.g., the diameter or the channel depth), and $v$ is the average velocity of the moving liquid. With this number we can obtain an impression of the flow behavior in microsystems. From empirical observations, physicists and engineers have found that Reynolds numbers larger than about 2300 correspond to what is called *turbulent flow*. Under this regime inertial forces are dominant, and this is the behavior we typically know from everyday life. As we can see from Eq. 3.4, large Reynolds numbers are attained at higher liquid densities, higher flow velocities, larger typical length scales, or lower viscosities.

The region in which the Reynolds number is between about 2000 and 3000 is called the regime of *transitional flow*, and the region in which it is below about 2000 is referred to as the *laminar* (or *creeping*) *flow* regime. Again, by examining Eq. 3.4 we can find that low Reynolds numbers are attained at lower velocities, smaller dimensions, smaller densities, or higher viscosities.

Let us consider an example:

Assume we have a microchannel with a circular cross section and a diameter of 100 μm. Further, let us use plain water as the liquid, which is being pumped through the microchannel at an average velocity of 1 cm s$^{-1}$. The values for the density and the viscosity can be found in Tabs 3.1 and 3.2, respectively. We can now calculate the Reynolds number for this example:

$$\text{Re} = \frac{0.998 \cdot 0.01 \cdot 1}{0.01} \approx 1 \tag{3.4b}$$

Clearly, we are in the laminar flow regime with this low Reynolds number. Conversely, we can ask how fast water has to be pumped through such a microchannel conduit before we reach the transitional flow regime. We choose to vary the velocity, because both the density and the viscosity are constant for water – in fact, both parameters are often combined into a parameter called kinematic viscosity, $\mu\,(\mu=\eta/\rho)$. The required pumping velocity thus is $v=20$ m s$^{-1}$. Things get even worse when we take different fluids, such as oil or air, for example, because oil has a higher viscosity and air has a very low density. For both oil and air, the kinematic viscosity is higher than for water and the velocities approach or exceed the speed of sound before the onset of turbulence.

Microchannels often do not have a cylindrical cross section, so the question arises: what is the best dimension to use for calculating the Reynolds number in a channel with, e.g., a trapezoidal cross section? To help in such situations the concept of hydraulic diameter was introduced. The hydraulic diameter, $D_h$, is defined as

$$D_h = \frac{4 \cdot A}{P_{\text{wet}}} \tag{3.5}$$

Here, $A$ is the cross sectional area and $P_{\text{wet}}$ is the wetted perimeter, which is all the perimeter that is in contact with the liquid. For a rectangular channel this corresponds to twice the width plus twice the height, whereas in a trapezoidal channel or a rounded rectangular channel it is a bit more complicated, but still readily available from standard geometric formulas. For a circular cross section Eq. 3.5 simplifies to $D_h=d$.

We have now looked at some basic parameters used to describe a liquid. How can these parameters be used to help describe the dynamics of a liquid in more detail? First, we have to decide whether we can still look at a liquid as a continuum instead of as an ensemble of 'individual' molecules. Does this view hold when going down into the micrometer domain? A simple calculation reveals that, in the typical volumes we deal with in microchannels (nanoliters to femtoliters), there are still very many liquid molecules present (on the order of $10^{10}$ in a femtoliter of water). Therefore, for almost all but the most extreme experimental conditions, the continuum assumption holds, which allows the scientist to rely on a vast body of work on fluid dynamics, established over more than a hundred years. The theoretical framework developed to analyze fluid flow is often referred to as Navier–Stokes formalism. It is built on the fundamental laws of conservation (mass, momentum, and energy), combining them with constitutive equations for fluids (governing viscosity and thermal conductivity) to arrive at a set of equations commonly referred to as Navier–Stokes equations. A more thorough description of these fundamental equations can be found in [1–3].

The Navier–Stokes equations contain more unknown parameters than equations, making complete analytical solution impossible. Typically, several boundary conditions and/or equations of state are adopted to help solve the Navier–Stokes equations under particular conditions. Among the most important and most-used boundary condition is the so-called *no-slip condition*, which states that the velocities at phase boundaries (i.e., wall–liquid) must be equal. This means that, for flow of a liquid inside a capillary or a channel, the fluid velocity at the wall must be zero ($v_{wall} = 0$). This has important implications for the velocity profile of such a flow.

An important solution to the Navier–Stokes equations is the Poiseuille (or Hagen–Poiseuille) flow, which applies when a pressure gradient is used to drive a liquid through a capillary or channel. For a capillary with a cylindrical cross section the following expression for the volume flow, $Q$, is found:

$$Q = \frac{\Delta V}{t} = \frac{\pi R^4}{8 \eta L} \Delta P \tag{3.6}$$

where $R$ is the radius of the capillary, $L$ is its length and $\Delta P$ is the pressure drop across this length (also called hydraulic pressure). The velocity profile, i.e., the velocities, $v(r)$, at different radial positions between the center ($r = 0$) and the wall ($r = R$) are found to follow

$$v(r) = (R^2 - r^2) \frac{\Delta P}{4 \eta L} \tag{3.7}$$

clearly describing a parabolic flow profile and also satisfying the no-slip condition.

The term, $8 \eta L / \pi R^4$, of which the reciprocal appears in Eq. 3.6, is also called the fluidic resistance. Note the dependency on $1/R^4$, which means that the fluidic resistance increases drastically as the channel dimensions are reduced. Consequently, higher and higher pressure drops are necessary to move liquid through smaller and smaller conduits. For channels with noncylindrical cross sections, expressions similar to those in Eq. 3.6 can be found, but with different terms for the fluidic resistance. Examples for common geometries are given in Tab. 3.3.

**Tab. 3.3** Fluidic resistances for some common channel geometries

| *Cross-section* | *Example* | *Fluidic resistance* |
|---|---|---|
| Circular | R | $8\, \eta L / \pi R^4$ |
| Rectangular (low aspect) | h  w | $12\, \eta L / w h^3$ |
| Square | a | $28.454\, \eta L / a^4$ |
| Regular triangle | a | $184.751\, \eta L / a^4$ |

Let us finish this part with a few important and useful relations and concepts, which can help during the design phase of microfluidic channel networks. The *continuity equation* deals with the behavior of flow in channels with changing cross sections. It simply states that the product of the cross-sectional area and the flow velocity is a constant, i.e.:

$$A_1 v_1 = A_2 v_2 = \text{const.} \tag{3.8}$$

The *Bernoulli equation*, on the other hand, looks at flows when pressure and height differences also play a role and is a direct application of the law of energy conservation, relating pressure, kinetic energy, and potential energy in the following way:

$$P + \frac{1}{2}\rho v^2 + \rho g h = \text{const.} \tag{3.9}$$

A knowledge of the value of $v$ (if only an average value) is useful because it can give an indication of the transit time of, e.g., a plug of chemicals or an ensemble of cells through a microfluidic channel network and therefore helps to assess whether there is enough time for a chemical reaction or mixing to take place (see also chapter 3.4). Both Eq. 3.8 and Eq. 3.9 are strictly only valid under idealized conditions (i.e., incompressible and nonviscous fluids and steady flows), but can still be helpful for overall estimates and assessments.

Finally, capillary and surface tension are forces to be reckoned with in microsystems. Indeed, surface tension directly determines how strong the capillary forces are in a microchannel, as qualitatively described for a capillary with a cylindrical cross section by Eq. 3.10:

$$F_{cap} = 2\pi r \gamma \cos \Theta \tag{3.10}$$

with $\gamma$: surface tension and $\Theta$: contact angle.

But beyond that, surface tension is also of great importance when dealing with bubbles and particulate contaminations in microchannels. A detailed treatment of this subject is, however, beyond the scope of this book.

## 3.2
## Transport Processes

### 3.2.1
### Types of Transport

In general, there are two very different types of transport within microfluidic systems, directed transport and statistical transport. The difference lies mainly in the nature of the driving agent behind the transport and manifests itself in the form of the transport. Directed transport is transport that is controlled by exerting work on the fluid. The work results in a volume flow of the fluid, where the flow can usually be characterized by a direction and a flow profile. The work is often generated mechanically by a pump or electrically by a voltage. Flow that is driven mechanically is called *pressure-driven* flow, and flow driven by a voltage is called *electroosmotic flow*.

Statistical transport differs from directed transport in that it is not a directly controlled transport. Instead, statistical transport is an entropy-driven transport, which means that transport occurs only if a fluid is more disordered after transport than before. A typical situation is at an interface between a liquid with a high concentration of one type of molecule and a liquid with zero concentration of the same molecule. This orderly situation leads spontaneously to a statistical transport of molecules from the side with high concentration to the side with zero concentration. Eventually, the concentration is equal in both liquids, the liquids are evenly mixed, and the situation is less ordered than before. This statistical transport is called *diffusion*.

It is rare that a transport process is purely directed or purely statistical. Much more common are mixtures of these types of transport. Mixed transports occur when a directed transport meets a gradient of some kind, which could be a temperature gradient or a concentration gradient. The molecules in the liquid follow the direction of the exerted outer flow, while at the same time the gradient is being equalized. A good example is forced-heat convection, in which there is a directed flow of molecules alongside a surface, with a heat-driven diffusion of molecules away from the surface.

### 3.2.1.1 Convection

The definition of *convection* in physics is heat transfer by mass transport. In everyday life convection is experienced where warm layers of fluids move into colder areas due to the difference in density that is caused by a temperature difference. This phenomenon is also called free convection or natural convection. Fluids can also be moved around with external forces to create a directed flow, which would then be called forced convection. In this section the heat transfer in forced convection is of secondary importance, since we concentrate on the transport of molecules.

There are several ways in which forced convection can generate directed flow, most of which have been mentioned above but are repeated here as a summary. If a microsystem is to be filled with its working liquid, but is still empty, then the first type of directed flow is a capillary flow, filling up the microsystem. In a similar fashion, other forces, such as gravity, a pressurized air bladder, or the centripetal forces in a spinning disk, create a single instant of pressure difference in the microsystem. Finally, there are mechanical and electroosmotic pumps, which are further discussed in section 4.3.

### 3.2.1.2  **Migration**

The directed transport of molecules in response to an electric field is called *migration*. In most cases, the moving molecules are ionized in a polar solvent such as water. These electrically charged molecules experience a coulombic force due to the electric field. The charged molecules first accelerate towards one of the electrodes, but then slow down and reach a terminal velocity, because they feel a drag caused by friction with the liquid. The coulomb force ($F$) is given by

$$F = qE \tag{3.11}$$

where $q$ is the charge on the molecule and $E$ the strength of the electric field. Once the molecules reach their terminal velocity, the coulombic force is balanced by a drag force, the Stokes force:

$$F = 6\pi\eta r v \tag{3.12}$$

where $\eta$ is the viscosity of the liquid, $r$ is the so-called hydrodynamic radius of the molecule, which indicates its size, and $v$ is the speed of the molecules. The terminal speed of the molecules is reached when both forces are equal and opposite, so that

$$qE = 6\pi\eta r v \tag{3.13}$$

From Eq. 3.13 the terminal speed is calculated as

$$v = \mu E \tag{3.14}$$

where $\mu$ is the mobility of the molecules, given by

$$\mu = \frac{q}{6\pi\eta r} \tag{3.15}$$

The relation in Eq. 3.14 is also true in 3 dimensions, in which both the electric field and the velocity are vectors. The electric field strength and the speed are the magnitudes of these vectors.

### 3.2.1.3  **Diffusion**

The process of diffusion is statistical by nature. In a liquid or gas, all molecules move in all directions as long as no external forces are applied. Each molecule moves in a certain direction for a certain time until it is hit by another molecule, whereupon it changes direction. Because the molecules are indistinguishable from one another, no net flow is observable. This also occurs when there are two types of molecules, say, large molecules surrounded by water molecules.

Diffusion occurs when there is a concentration gradient of one kind of molecule within a fluid. For example, when a concentration gradient of large molecules in water is present, one can notice net movement of large molecules away from areas of high concentration towards areas of low concentration. The main reason for this is that there are more large molecules in areas of high concentration than in areas of low concentration, so that many molecules move randomly in one direction, but only a few molecules move randomly backwards.

The statistical movement of a single molecule in a fluid can be described as a random walk. The movement is characterized by the Einstein–Smoluchowski relation:

$$x = \sqrt{2Dt} \tag{3.16}$$

where $x$ is the average distance moved after an elapsed time $t$ between molecule collisions and $D$ is a diffusion constant that is characteristic for the given molecule. The most important conclusion that can be deduced from Eq. 3.16 is that the moved distance is proportional only to the square root of time, in contrast to directed movement, where the distance is directly proportional to time. This finding can be attributed to the fact that the molecules move randomly and not in a directed way. Eq. 3.16 also shows that a larger diffusion constant means faster movement. In general, it is true that the larger a molecule is, the smaller is its diffusion constant.

Eq. 3.16 allows us to estimate the time $t_{cross}$ it takes a molecule to cross half the channel width $W$ of the main channel of a T-junction:

$$t_{cross} = \frac{W^2}{8D} \tag{3.17}$$

This means that the diffusion dimension, here the channel width, of a microchannel has a large influence on the mixing of liquids in a channel. This is discussed in greater detail in sections 3.4 and 4.5.

The overall effect of the random walk of all molecules can be described as temporal and spatial changes in concentration. These changes are related to each other by Fick's second law of diffusion, which can be derived from Fick's first law of diffusion and the continuity equation. To understand Fick's first law of diffusion, which is often just called Fick's law, the particle flux $j$ is introduced. The flux is the number of molecules crossing a certain area $A$ during a time span $t$ and is given by

$$j = \frac{N}{At} = cv \tag{3.18}$$

where $c$ is the molecules' concentration and $v$ their average velocity. This equation takes into account that the molecules cross a certain distance within the time span, which can be multiplied on the nominator and denominator of the fraction.

The observation that molecules travel from areas of high molecule concentration to areas of low concentration is summarized in Fick's law as

$$j = -D\frac{\partial c}{\partial x} \tag{3.19}$$

The proportionality constant is the diffusion constant, as in Eq. 3.16. Eq. 3.19 indicates that the molecule flux is strongest where the concentration gradient is steepest. Fick's law is a empirical law, which for all practical purposes approximates the underlying statistical processes well.

In section 3.2 the continuity equation is introduced in its simplest appearance, in which the product $Av$ is constant for a flow in a microchannel of cross section $A$ and average velocity $v$ (Eq. 3.8). This form of the continuity equation was arrived at by assuming that the concentration of the molecules being considered was constant, which was evidently the case, because flow of only one type of liquid was looked at. It is also possible to formulate a continuity equation for molecules dissolved in a solvent, where the possibility of a change in concentration must be taken into account. The continuity equation is one manifestation of the principle of conservation of mass. As far as molecules are concerned, it can be stated that when molecules leave a test volume, then the number of molecules in the test volume is correspondingly lower. Mathematically, the continuity equation is given as

$$\frac{\partial c}{\partial t} = -\frac{\partial j}{\partial x} \tag{3.20}$$

The continuity equation and Fick's law can be combined into Fick's second law of diffusion, which can be expressed in one dimension as

$$\frac{\partial c}{\partial t} = D\frac{\partial^2 c}{\partial x^2} \tag{3.21}$$

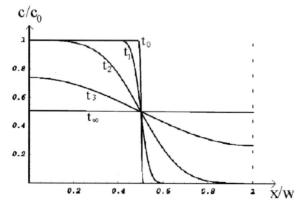

**Fig. 3.2** Development of a diffusion front. The graphs show the solution of the one-dimensional diffusion equation, Eq. 3.21. The initial condition was zero concentration for $x$ values greater than 0.5 W and constant concentration for $x$ values smaller than 0.5 W.

If an area of high concentration directly borders an area of zero concentration, then the sharp border fades slowly over time. Fig. 3.2 shows the fading process, calculated from the solution to Eq. 3.21 for a typical diffusion constant.

**Temperature dependence of diffusion**

Interestingly, a relation between the macroscopically observed diffusion behavior in the form of the diffusion constant $D$ and a microscopic property of the molecules in the form of the hydrodynamic radius $r$ can be stated. This is the Stokes–Einstein relation:

$$D = \frac{kT}{f} , f = 6\pi\eta r \tag{3.22}$$

where $f$ is a frictional constant known from the Stokes equation (Eq. 3.12), $k$ is the Boltzmann constant, and $T$ is the temperature. Eq. 3.22 shows the temperature dependence of the diffusion constant and allows us to estimate the diffusion constant of molecules when the hydrodynamic radius is known.

**Relating convection and diffusion – Peclét number**

Directed and statistical flow in microsystems are discussed in previous sections (see 3.2.1.1 and 3.2.1.3). Under different circumstances and for different geometries of microfluidic systems, one or the other of the flow types dominates, or both flow types might be of equal importance. To evaluate the various flow situations, we can examine the ratio between the mass transport due to directed flow and that due to diffusion. This ratio is a dimensionless number, the Peclét number ($P_e$), given as

$$P_e = \frac{vd}{D} \tag{3.23}$$

where $d$ is a characteristic length of the microfluidic system.

As dimensionless numbers, both the Peclét number and Reynolds number are typical of the field of hydrodynamics and allow making a statement about the importance of a given type of flow phenomenon in a system. Apart from the Reynolds and Peclét numbers, there are many more dimensionless numbers used to evaluate phenomena such as free heat convection and capillarity. In this section, however, the Peclét number is the most important number to consider. In a microfluidic system, the most important geometric dimensions, the operating conditions and the properties of the involved molecules are often known. Under most circumstances, the length and width of a microchannel, the average speed of the working fluid as well as the diffusion constants of the molecules of interest are therefore known. The Peclét number of the microchannel under question can be calculated with this information and the effect of diffusion compared to the directed flow can be evaluated. The gained information is important in the design of microfluidic systems that need to keep control of diffusion, such as chemical separation systems.

### 3.2.1.4 **Dispersion**

#### Band broadening

Diffusion limits the resolution of an analytical method called flow injection analysis, which is discussed in more detail in section 10.3. An essential part of flow injection analysis is the insertion of a sample plug into a flowing buffer stream. The sample plug contains a high concentration of certain molecules and has relatively sharp boundaries with the buffer stream. While the plug moves along a microchannel, the concentration boundaries become vaguer due to diffusion of molecules or to the fact that some molecules must travel longer distances than others, such as in the race-track effect. The widening of the sample plug is generally called *dispersion*. When a signal is detected from the analysis, then the widening is also known as band broadening.

#### Taylor dispersion

Taylor dispersion is a superposition phenomenon. In addition to the band broadening caused by diffusion, there is band broadening caused by the parabolic velocity flow profile, which occurs when the flow is pressure driven. Because the liquid in the center of the microchannel flows faster than the liquid at the edges, the central parts of the sample plug are gradually separated from the outer parts. In this case of velocity-difference-driven band broadening, diffusion has a positive influence on the resolution of flow injection analysis. The reason for this is that some molecules diffuse from low-velocity areas of the sample plug to high-velocity areas and vice versa. The overall effect is a tighter sample plug than in the absence of diffusion, as shown in Fig. 3.3. This superposition of direct movement and diffusion is known as Taylor dispersion.

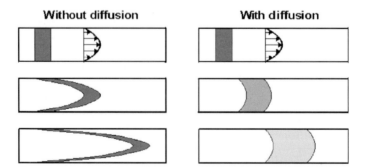

**Fig. 3.3** Taylor dispersion. The figure shows concentration levels within a microchannel. The figure is a result of numerical simulations of the superposition of pressure-driven flow along a microchannel and diffusion across the same channel (drawing courtesy of Flemming Rytter Hansen, MIC).

**3.3**
**System Design**

The Peclét number is useful in designing a microsystem for flow injection analysis. Knowing the Peclét number of a microchannel and its dimensions, the designer can distinguish between two major situations. First, if the Peclét number is much smaller than 1, then diffusion dominates the microfluidic flow, and directed flow is of secondary importance. Second, if the Peclét number is much larger than 1, the molecules of interest flow mainly according to the externally applied driving force, and diffusion has only a minor influence. In microsystems, the flow velocities are usually comparatively small. The crucial variable that determines the Peclét number is therefore the channel lengths $d$. For long enough channels, the Peclét number is always larger than 1, and the flow is consequently directed.

Another distinction between different types of flow involving the Peclét number is the difference between Taylor flow and purely advective flow. If the Peclét number is much smaller than the length-to-width ratio of the microchannel ($d/w$), then Taylor dispersion is observed. If the Peclét number is much larger than the length-to-width ratio however, then diffusion is not the main agent of dispersion. The two situations are different, because a relatively narrow microchannel allows transverse diffusion to play a significant role, which does not happen in relatively broad channels. Once the designers have decided upon the length, width, and flow speed of the separation system, they can adjust the microchannel volume by choosing the depth of the channels.

For further information on this section see [1, 8–10].

**3.3.1**
**Laminar Flow and Diffusion in Action**

From the previous two sections it is clear that the two most important phenomena we face in microfluidic devices are the almost exclusive presence of *laminar flow* and the dependence on *diffusion* as the major available transport mechanism. We should stress, however, that turbulent flow may occur under certain rare conditions, for example, when geometries change rapidly and do not allow the flow to reach a steady state before the next change in geometry. For such situations,

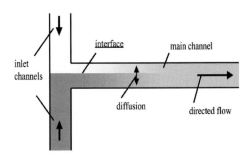

**Fig. 3.4** A T junction where two liquids meet and flow downstream side by side (laminar flow), slowly mixing only by diffusion.

modeling and simulation tools are becoming increasingly important and useful for the design and evaluation of microfluidic devices. Please see chapter 5 for more on this subject.

Let us now look at the possibilities for using laminar flow and diffusion in a constructive and creative way to implement functional elements and processes inside a microfluidic device. First, we take the simple example of a T or Y junction, where two different streams meet (Fig. 3.4). Contrary to what we would experience with macroscopic streams, the two liquids meeting in such a fashion on a microdevice do not immediately mix by turbulence, but move alongside each other down the main channel. The only lateral transport mechanism that is available is diffusion. If we assume that the diffusion coefficients are given by the nature of the molecules, the solvents used, and the external experimental conditions (temperature, etc.), what degrees of freedom do we have to influence the behavior of such a microfluidic system? The main design parameters are the width of the main channel, where half the width corresponds to the worst-case diffusion distance, and the linear (average) flow velocity, which determines the contact time between the two streams and therefore translates into a minimum length of the main channel at any given flow velocity. Let us look at an example: stream A contains a solution of molecules with diffusion coefficient $D=10^{-5}$ cm$^2$/s in water, and stream B contains pure water. The main channel has a width of 200 µm, and the average linear flow rate is $v=1$ cm s$^{-1}$. How much time does it take to achieve complete mixing?

The largest distance a molecule has to diffuse from stream A into stream B is half the width of the channel, i.e., 100 µm in this example. With Eq. 3.16 we can calculate the time it takes to diffuse this distance to be $t_{diff}=5$ s. With the given linear flow velocity, a main channel length of 5 cm is required to achieve mixing before the liquids leave this section. However, because of the square root dependence (Eq. 3.16) we can gain a lot by reducing the diffusion distance, i.e., the channel width (at the same time increasing the channel depth so as not to increase the overall fluidic resistance). For a channel width of 90 µm the time for complete mixing by diffusion is reduced to 1 s, allowing mixing to be achieved within 1 cm of channel length. Thus, reducing the diffusion distance saves time and microchip real estate and reduces problems of increased fluidic resistances because of too-long channel segments ($R \propto L$, see Tab. 3.3). By the same approach we can select conditions (channel width, flow velocity, or channel length), in which hardly any mixing occurs and in which the two streams flow almost unperturbed side by side down the channel. How this can be used favorably is discussed below.

Mixing is one of the challenges in microfluidic devices. All passive mixing (i.e., without external energy input) relies primarily on diffusion. Strategies to improve mixing include splitting streams into smaller streams and folding and relaminating these streams again and again, thereby again minimizing the diffusion distances. Using an array of holes to inject plumes of liquid A into liquid B also gives large interaction surfaces and short diffusion distances. Introducing additional (lateral) transport by means of special topographies has also been suggested and tested. A more detailed description of the most common strategies for passive and active mixing is given in chapter 4.3.

**Fig. 3.5** A 'filtering' device utilizing laminar flow and diffusion (figure adapted from ref. [11]).

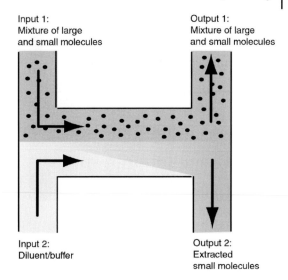

Input 1:
Mixture of large
and small molecules

Output 1:
Mixture of large
and small molecules

Input 2:
Diluent/buffer

Output 2:
Extracted
small molecules

With the knowledge of how diffusion time and contact time can be influenced by proper design of a microfluidic layout, we can easily devise ways to make use of this know-how. A simple yet effective example is the so-called H filter (Fig. 3.5). It has two inlet channels, a section where the two liquids are laminated and flow side by side, and then two exit channels. If liquid A contains a mixture of two different molecular species, one with a small diffusion coefficient and one with a rather large diffusion coefficient, and liquid B contains pure solvent, the following happens at the H crossbar where both liquids are laminated: depending on the width of the crossbar and the flow velocity, only a very tiny fraction of the molecules with a small diffusion coefficient diffuse into liquid B in the given contact time. On the other hand, a much larger fraction of the molecules with a large diffusion coefficient can diffuse into liquid B. Thus, this arrangement works like a filter or an extractor, selecting between two types of molecules in a continuous way. Building on this rather simple principle, several more advanced layouts and uses are feasible in microfluidic devices.

Another very popular design is a layout (Fig. 3.6) [12] that features three inlet channels merging into a common channel, and then, depending on the intended use, one to several exit channels. It is not crucial how the channels are arranged geometrically, whether merging at an acute angle or in a right-angle geometry. On the inlet side the liquid of interest typically enters the middle channel, and two flanking liquids can be used to guide the center liquid and even focus or compress it slightly – all by using the fact that only laminar flow occurs in microsystems. Control is achieved fairly easily by merely adjusting the relative or absolute flow rates (volume flow) of all the flows involved. For example, to focus or narrow the center flow to a greater or lesser extent, the laminating flow rates can be increased or decreased. In this way it was even possible to increase the concentration of a chemical species present in the center flow, right at the point where the

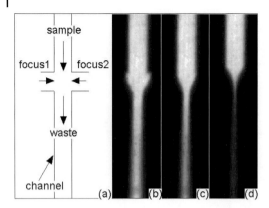

**Fig. 3.6** Flow focusing by using two laminar side flows (a) and various flow ratios (b–d) (reprinted with permission from [12]. Copyright (1997) American Chemical Society).

two focusing side streams join the center stream. Another scenario has cells in the center flow. Due to the laminating and slightly focusing effect of the side flows, cells are confined to a narrow, well-defined part of the channel, from which they move downstream as single cells, one after another. The laminating also assures that all cells pass a detection window, which is typically smaller than the channel width. Techniques for counting, sizing, and sorting cells make great use of these behaviors in microfabricated channels. An additional challenge is the velocity distribution of the cells as caused by a parabolic flow profile. Cells closer to the walls have lower velocity than cells in the center of the channel. Lamination can help focus cells into a narrow corridor within the channel, thus making sure that they all have similar velocities. One has to be careful, however, not to forget that these systems are three-dimensional and that the flow focusing described here works only two-dimensionally. Therefore, cells can still show different velocities because their positions with respect to the bottom and top of the channel are not defined by laminating steering flows. To improve this situation new designs with truly three-dimensional lamination have been proposed, fabricated, and tested. Again, this is an example for which both simulation tools and experimental techniques can give extremely helpful information on actual flow behavior in microchannels. Section 3.4 shows some results on blood flow in microchannels obtained with an experimental technique called particle image velocimetry (PIV).

If we do not adjust the two laminating flow rates simultaneously, but independently, we do not get focusing, but instead have a tool to control the lateral position of the center flow within the channel. It is thus possible to selectively guide the center flow to different exit channels (Fig. 3.7). This is also used for sorting cells, for example, because the center stream can be deflected for a short period of time into a collection channel whenever a cell of interest is detected, but at all other times the center stream is diverted directly to waste. Similarly, streams of reagents can be guided to different channels to initiate different reactions, or fractions eluting from a separation unit can be collected in different channels.

Let us look at more examples of how to make good use of lamination and diffusion in microsystems. Often, chemicals need to be available at different concentra-

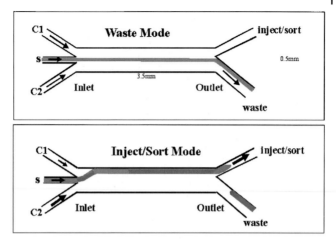

**Fig. 3.7** Functioning principle of a flow-sorting device using laminar flow (figure courtesy Anders Wolff, MIC).

tions, e.g., to calibrate a sensor or to adjust pH levels according to the input solution. On-chip diluters, which can provide different concentrations starting from a very concentrated solution, are therefore necessary. One example of a serial mixer/diluter is shown in Fig. 3.8 [13]. On the left side of the layout sample (the concentrated solution) and pure buffer or solvent are brought together. (Undiluted sample can be shunted off at this point.) The sample and buffer are then given a certain amount of channel length in which to mix by diffusion, resulting in dilution of the sample. In Fig. 3.8, this part is the bent sections between intersections. The mixing channel is bent only to save chip real estate and to give the two liquids an appropriate amount of channel length and thus contact time. At the next intersection, the diluted sample is partly shunted off for use, and partly laminated with more pure buffer for further dilution. This procedure is then repeated several times, producing more-and-more-dilute sample solutions and making them available for further use in a different part of the chip.

By using a more elaborate design, but basically the same principles, complex gradients can be designed and created (Fig. 3.9) [14]. Here, careful, repeated splitting and recombining flows at T-junctions and then relaminating them again in a common channel creates complex concentration distributions across the common channel. Several applications can benefit from this, e.g., separation techniques requiring defined pH gradients across the channel width.

To finish this section, let us look at an exciting example of how to use lamination and diffusion to create truly tiny structures within already microfabricated channels. As two liquids are laminated a reaction can take place at the interface between them. The 'reaction zone' is limited to a small region around the interface, defined by diffusion. In this way it was possible to demonstrate, e.g., the etching of a small trench into the bottom of a microchannel (Fig. 3.10) [15]. To achieve this, a solution of potassium fluoride and a solution of hydrochloric acid

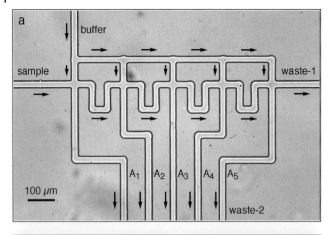

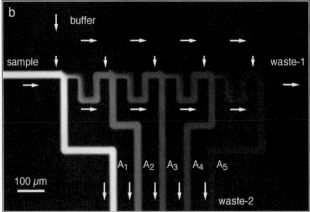

**Fig. 3.8** A serial mixing device delivering several degrees of dilution from two inputs (sample and buffer); a) microphotograph of device; b) serial dilution of a fluorescent dye (reprinted in part with permission from [13]. Copyright (1999) American Chemical Society).

were laminated, creating hydrofluoric acid at the interface. This acid etched the substrate material, producing a trench at the lateral position of the interface between the two solutions. In a similar way, the researchers could deposit a tiny silver wire at the interface of a silver salt solution and a reducing solution (Fig. 3.11) [15]. Beyond these mere demonstrations, they produced a miniaturized electrochemical cell within a channel with the approaches described. They first selectively etched a gold patch on the bottom of a channel by laminating a gold etching solution with two buffer streams, thereby protecting part of the gold, which was to become the working and counter electrodes. Then, again by laminating two suitable solutions they deposited a silver wire in between the separated gold patches, thereby creating a reference electrode and finishing the electrochemical detection cell (Fig. 3.12) [15].

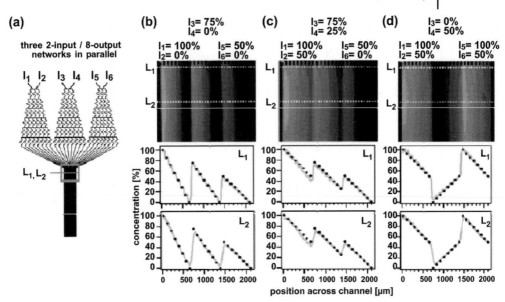

**Fig. 3.9** Complex gradients across a microchannel created by splitting and combining flows and making use of lamination and diffusive mixing (reprinted with permission from [14]. Copyright (2001) American Chemical Society).

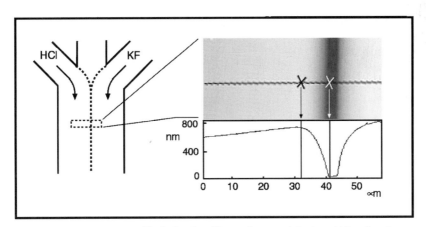

**Fig. 3.10** In situ generation of hydrofluoric acid to etch a trench in the middle of a microchannel (reprinted with permission from [15]. Copyright (1999) American Association for the Advancement of Science).

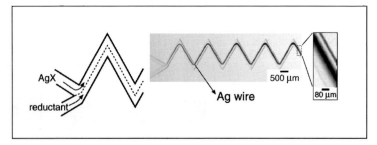

**Fig. 3.11** In situ creation of a silver wire inside a microchannel (reprinted with permission from [15]. Copyright (1999) American Association for the Advancement of Science).

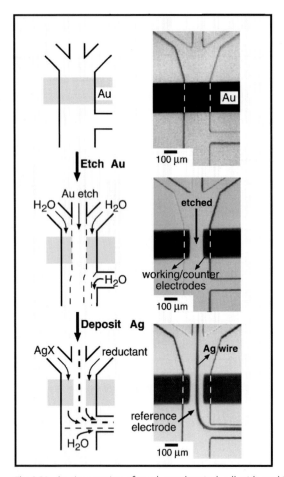

**Fig. 3.12** In situ creation of an electrochemical cell with working, counter, and reference electrodes inside a microchannel (reprinted with permission from [15]. Copyright (1999) American Association for the Advancement of Science).

**3.4**
**An Application: Biological Fluids**

One major market for microfluidic devices will be the health sector, where micro-chips will be used for diagnostics, monitoring, and drug delivery. Thus, studies on whether biological matrices and microfabricated devices are compatible and work together well are of crucial importance. One of the most interesting and challenging biological matrices is, of course, blood. Handling of small amounts of blood is tricky, its compatibility with many materials used in micromachining is limited, and there is not much knowledge about the flow behavior of blood in micochannels, even though every day blood rushes through the tiniest capillaries in our bodies. This last challenge is mainly due to the non-Newtonian behavior of blood. Although a challenge to the researcher, it is precisely this behavior that allows blood to be pumped into smaller and smaller capillaries as it moves through our body, because the viscosity changes as the shear strain is increased when entering a tinier capillary. The non-Newtonian behavior stems mostly from the fact that blood is composed of plasma and corpuscles, and these corpuscles can change shape and squeeze into tubing seemingly too small for them. However, this behavior also makes it much more difficult to approach the problem with the help of computer simulation tools. Consequently, experimental investigations of the flow behavior of blood in microchannels are urgently needed. One promising possibility is micro-particle image velocimetry (micro-PIV). In this technique, beads with diameters between a few hundred nanometers and a few micrometers are added to the interesting streams and visualized by microscopy. Photographs taken at short intervals can be used to correlate the movement of the beads and calculate backwards to the velocity distribution in the observed region of the micro-

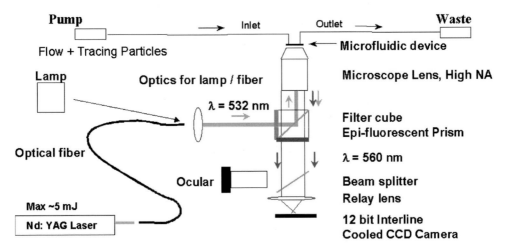

**Fig. 3.13** Schematic of a microparticle image velocimetry (micro-PIV) setup to determine velocity fields within microfluidic channels (figure courtesy of Carsten Westergaard, Dantec).

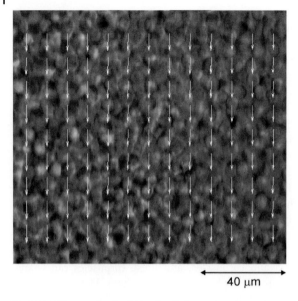

40 μm

**Fig. 3.14** Micrograph of red blood cells in a microchannel with overlaid velocity vectors (arrows) obtained from PIV measurements (picture courtesy of Lennart Bitsch, MIC).

channel. With blood, one does not need to seed the solutions with beads, but instead can directly see the leukocytes or erythrocytes in the blood.

Fig. 3.13 shows a schematic of a micro-PIV setup, and Fig. 3.14 shows a photograph of blood in a microchannel. From such photographs it is possible, by using appropriate algorithms, to determine the velocity field in the observation area (arrows in Fig. 3.14). Straight channels can be used to calibrate the PIV setup and to get a feel for the experimental challenges for measuring blood directly. Later, more advanced experiments can be performed in structures that are more interesting and promise to give new insights into the flow behavior of blood. Such structures include sharp bends and restrictions in the flow path. With the knowledge acquired from experiments like these, together with input from theoretical models, it will be possible to design microdevices for handling small amounts of blood to perform medical and chemical analyses.

## 3.5
## References

1 PROBSTEIN, R. F. *Physicochemical Hydrodynamics: An Introduction*, 2nd edit., Wiley: New York, 1994.

2 KARNIADAKIS, G. E.; BESKOK, A. *Microflows*, Springer: New York, 2002.

3 SHERMAN, F. S. *Viscous Flow*, McGraw-Hill: New York, 1990.

4 BATCHELOR, G. K. *An Introduction to Fluid Dynamics*, Cambridge University Press: Cambridge, UK, 2000.

5 NGUYEN, N.-T.; WERELEY, S.T. *Fundamentals and Applications of Microfluidics*, Artech House: Boston, 2002.

6 *CRC Handbook of Chemistry and Physics*; CRC Press, Boca Raton, FL, 1985.

7 SERWAY, R.A. *Physics for Scientists and Engineers*, 3rd edit., Saunders College Publishing: Philadelphia, 1990.

8 ATKINS, P.W. *Physical Chemistry*, 5 edit., Oxford University Press, 1994.

9 INCROPERA, F.P.; DEWITT, D.P. *Introduction to Heat Transfer*, Wiley: New York, 2002.

10 MOLHO, J.I.; HERR, A.E.; MOSIER, B.P.; SANTIAGO, J.G.; KENNY, T.W.; BRENNEN, R.A.; GORDON, G.B.; MOHAMMADI, B. *Analytical Chemistry* 2001, *73*, 1350–1360.

11 KOVACS, G.T.A. *Micromachined transducers sourcebook*, McGraw-Hill: New York, 1998.

12 JACOBSON, S.C.; RAMSEY, J.M. *Analytical Chemistry* 1997, *69*, 3212–3217.

13 JACOBSON, S.C.; MCKNIGHT, T.E.; RAMSEY, J.M. *Analytical Chemistry* 1999, *71*, 4455–4459.

14 DERTINGER, S.K.W.; CHIU, D.T.; JEON, N.L.; WHITESIDES, G.M. *Analytical Chemistry* 2001, *73*, 1240–1246.

15 KENIS, P.J.A.; ISMAGILOV, R.F.; WHITESIDES, G.M. *Science* 1999, *285*, 83–85.

# 4
# Microfluidics – Components

Jörg P. Kutter, Klaus Bo Mogensen, Henning Klank, and Oliver Geschke

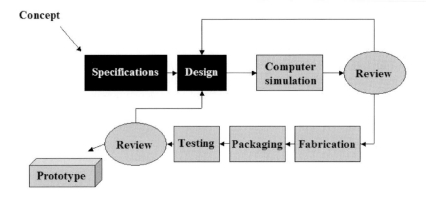

## 4.1
## Valves and Pumps

Among the most important fluid-handling elements are pumps and valves. Pumps are devices to set fluids into motion, and valves are designed to control this motion and, for example, define a preferred direction for the motion of the fluid. The engineering community has developed a wealth of pumps and valves based on a large variety of principles. Many of these principles and implementations can also be used in the microdomain, but additionally new designs have appeared making clever use of the different behavior of liquids and mechanics on small scales. It is far beyond the scope of this book to cover all proposed and tested micropump and microvalve designs. In fact, new designs appear almost daily. A good overview of many pump and valve principles adapted or developed for microfluidic devices can be found in, e.g., [1–4]. In this section we look at but a few selected examples of how pumping and valving are implemented in microsystems. A special section (4.4) is dedicated to pumping by electroosmosis, because this is a unique and widely applied principle deserving special attention. The remainder of this section deals with pumps generating so-called hydrodynamic flow, typically based on a pressure difference between the inlet and outlet of a channel.

*Microsystem Engineering of Lab-on-a-chip Devices*
O. Geschke, H. Klank, P. Telleman
Copyright © 2004 Wiley-VCH Verlag GmbH & Co. KGaA, Weinheim
ISBN: 3-527-30733-8

A valve is a device within a fluidic system in which flow is allowed in one direction but suppressed in the opposite direction, thus introducing directionality into the flow. Valves are often classified by whether they work by themselves (i.e., by utilizing energy from the flow) or whether they need external energy to function. The former are called passive or check valves, whereas the latter are called active valves. The wish-list of ideal characteristics that a valve should have includes zero leakage (when it's closed it's closed, when it's open it's open), zero power consumption (obviously not true for active valves), zero dead volume (should not introduce extra volumes, which negatively affect the performance), infinite differential pressure capability (a little bit of extra pressure from one side should open it, a little bit of extra pressure from the other side should close it), zero response time (no delays), insensitivity to particulate contamination, ability to operate with any fluid, and so on. Unfortunately, no single valve meets all these requirements, in other words, all valves have some flaws, which need to be taken into account when operating them.

An example of a passive valve is the cantilever type (Fig. 4.1), here, a thin strip of silicon that bends when enough pressure is applied from one side. From the other direction bending is restricted by the valve seat. Depending on the exact design and operating conditions, these kinds of valves are more or less leaky. However, they are relatively simple to make, do not require external energy for operation, and have a fairly fast response time. There are also passive valves, which do not rely on mechanical action, but rather make use of forces such as surface tension. These types of valves are typically one-time-use or burst valves, because their functioning depends on an air–liquid interface, which is typically present only when first filling a microchip. A simple implementation of such a valve is a restriction in the channel. As the air-liquid interface approaches the restriction the

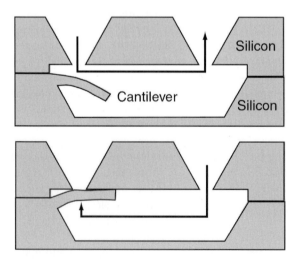

**Fig. 4.1** A check valve using a cantilever as a valve flap (adapted from [1]).

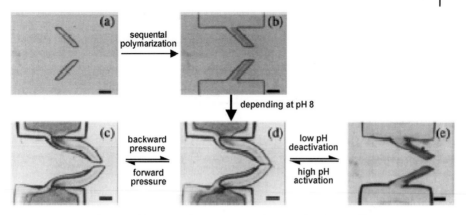

**Fig. 4.2** Example of a valve fabricated of pH-sensitive hydrogels, which have different swelling behaviors depending on the pH of the solution (reprinted with permission from [5]. Copyright (2001) American Institute of Physics).

liquid cannot penetrate into the next channel segment because of the increased surface tension around the restriction. Only when a higher pressure is applied will the liquid break or burst through the restriction into the next section of the channel network. Similarly, defined patches on the channel bottom, which are chemically treated to have a hydrophobic surface can prevent aqueous liquids from flowing until sufficient pressure is applied to overcome the surface tension. Recently, another type of (basically passive) valve has appeared, based on hydrogels. Hydrogels are materials that can change some physical parameter (typically, their volume) based on some physical or chemical stimulus. One example uses small strips inside a channel made of different hydrogels, one of which is pH sensitive (Fig. 4.2) [5, 6]. Depending on the pH of the solution flowing through the channel, the hydrogel swells or stays in the shrunken state. If it swells the flaps block the passage of solution in a certain direction. Because the valve in a way 'senses' the solution and acts accordingly, these valves have also been called intelligent or adaptive valves.

Active valves require an actuator to provide a mechanical action, which moves a part of the valve to close or open the flow passage. Valves can be designed to be 'normally open' or 'normally closed', meaning that this is the state they are in without actuation. For reaching and maintaining the opposite state, energy needs to be applied. A few designs have two stable states (open and closed) and need energy only for transition between the two states (so-called bistable valves). The actuation principles used are various, including pneumatic (using compressed air), thermopneumatic (using heated fluids), piezoelectric (using special materials that expand when an electrical voltage is applied), electrostatic (using electrical attraction or repulsion), shape memory alloy (metals that change shape under temperature changes but 'remember' and revert to their original shape), and electromagnetic actuation, to name but a few. Actuators are typically characterized by four

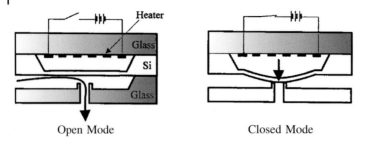

Open Mode            Closed Mode

**Fig. 4.3**   An active valve using a heating chamber with a working fluid and a diaphragm (reprinted with permission from [3]. Copyright (1998) Springer Verlag).

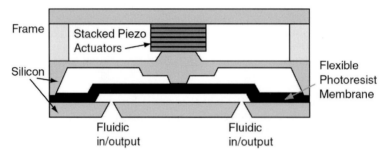

**Fig. 4.4**   An active valve having a piezoelectric stack as the actuation element (adapted from [1]).

criteria: the pressure they can build up, the stroke displacement they generate, their response time, and their reliability. Again, some actuators perform better in one category but less well in another. For example, piezoelectric stacks can produce large pressures, but not very impressive strokes. Electromagnetic actuators, on the other hand, can give large displacements but only relatively small pressures. A valve with a thermopneumatic actuator is depicted in Fig. 4.3.

A range of heating elements can heat a working fluid in a chamber on top of a diaphragm (50 µm thick silicon). The expansion of the fluid bends the diaphragm, which presses onto a valve seat sealing off the liquid flow. A different design with a piezoelectric stack as the actuator is shown in Fig. 4.4. Here, a thin, flexible photoresist membrane is included in the design to provide a tighter seal, i.e., less leakage. Softer gasket materials are supposed to be better suited to handling particulate contamination, which can prevent proper functioning of a valve especially when only hard materials are used.

Many different valves are available and imaginable, all using different designs, materials, and actuation principles. There is no optimum valve, but instead valves have to be chosen carefully depending on the intended use of the microsystem. Questions such as: what kind of solution is being moved through the system, i.e., comes in contact with the valves?, how often do we need the valves (many times per minute, only a few times in total)?, can we afford to spend a lot of energy on actuating the valves? and so on have to be considered before designing or choosing a valve. Often, microfabricated valves still have too many flaws or drawbacks and researchers revert to off-chip, conventional miniaturized valves. The same can apply for pumps.

In its most common version, a microfabricated pump consists of a pump chamber and two valves: an inlet valve and an outlet valve. The pump chamber is mostly a combination of a diaphragm/membrane and an actuator to displace the membrane or diaphragm. This displacement, together with the action of the valves, results in alternating flux into and out of the chamber, i.e., pumping. The valves used can be any of the aforementioned valves, passive or active, as long as they can be used repeatedly over a long period of time. There are also valveless pumps as discussed below.

The requirements for a microfabricated pump can be as manifold as its intended uses. A typical list of requirements for a microfabricated pump to be used in a microsystem for analyzing wastewater contained the following points [7]:

- as little pulsation as possible
- flow rate controllable over a certain range (e.g., 0.25–10 $\mu$L min$^{-1}$)
- leak rates under 1 nL min$^{-1}$ at counterpressures of 1–10 kPa
- sensitivity to counterpressures of 1–10% kPa$^{-1}$
- flow rate precision over several days within 1%–3%
- resistance to aggressive chemicals over a longer period of time at elevated temperatures
- production costs as low as possible

Such requirements are made to assure accurate analytical results, reliable operation, and economic viability of a microsystem utilizing such a pump. At the same time, many of these requirements are extremely hard to meet, and meeting some of them often means seriously compromising other requirements. For instance, to be able to withstand aggressive chemicals (pH 11–12) over a long period at higher temperatures (about 40–50 °C) almost prohibits the use of silicon as a material for the microfabricated pump. Alternative materials need to be investigated, for which only a limited range of machining possibilities exist, and which have different mechanical properties (see chapter 7 and chapter 8). Such materials might also prevent the application of already developed and tested silicon-based actuators.

Alternatively, protective coatings could be applied, making additional processing steps necessary and also affecting the mechanical performance of the device.

As mentioned above, microfabricated pumps employ the same type of actuators as active valves. The combination of actuators chosen for the pumping chamber and the valves, respectively, determines the performance of the overall pump, at least with respect to generated pressure, stroke displacement, response time, and reliability. Additionally, the achievable flow rate and the amount of pulsation are criteria for microfabricated pumps. In a pump with piezoelectric actuation of the pump chamber and two ring-type passive valves (Fig. 4.5), even though piezocrystals can be driven with a very high frequency, the employed type of passive valves do not respond extremely fast and therefore limit the range of possible driving fre-

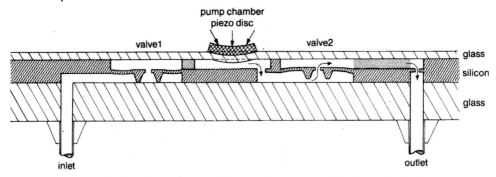

**Fig. 4.5** A relatively simple pump design involving a pump chamber with a diaphragm, a piezo-electric actuator, and two ring-type passive valves [4] (courtesy H. van Lintel, EPFL).

quencies. Cantilever-type passive valves have a faster response, but provide a less tight seal and are more prone to leaking. A pump design that works entirely without valves, thereby reducing the amount of moving, and potentially failing, parts is shown in Fig. 4.6. Here, a pump chamber with a membrane is actuated as in many other pump designs. However, instead of inlet and outlet valves, asymmetric restrictions separate the pump chamber from the rest of the channel system. Depending on the exact geometry of these asymmetric restrictions, fluidic resistance in the two possible directions differs, thereby rectifying the flow into a preferred direction. The asymmetric restrictions function as either a diffuser or a nozzle, depending on whether flow is into or out of the chamber, which is why they are also referred to as diffuser–nozzle elements. Obviously, a micropump with diffuser–nozzle elements always has a relatively high degree of leakage, but is potentially more reliable because it has fewer mechanical parts.

All of the many designs of microfabricated pumps with actuators suffer from delivering pulsating flow, which is unavoidable because all actuators use a reciprocating movement. An elaborate design in which several of these pumps work together out of phase could improve the situation somewhat.

However, other (simpler) approaches to micropumps can deliver a steady, non-pulsating flow, much like a conventional syringe pump. One possibility is to use a constant-pressure gas reservoir, which presses on the liquid and displaces it. The gas reservoir could be pressurized by a microfabricated (pulsating) gas pump and regulated by valves and pressure sensors. One of the additional advantages of such a design is that the potentially corrosive liquid never comes in contact with the micropump itself. Pressure differences between two ends of a channel can also be generated by simply creating a difference in the liquid filling level or by placing a small and a large droplet at the ends and utilizing the larger surface tension of the smaller droplet. If channels need to be filled only once (also called priming) capillary action can be exploited to move liquids into channel networks. Even though the above-mentioned approaches are relatively simple, more advanced designs are required if flows of constant magnitude are desired.

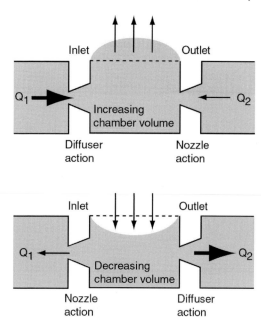

**Fig. 4.6** Diffuser–nozzle designs to direct flow. Any of the mentioned actuation principles can be used to actuate the pump (adapted from [1]).

Another very interesting way to provide nonpulsating flow is to use centrifugal force. Imagine microchannels fabricated onto a CD-like disc and arranged in a mostly radial pattern (Fig. 4.7). If the disc is set spinning, centrifugal force moves the liquid inside the channels from the center of the disc towards its rim. Valving in this case is achieved with, e.g., burst valves or hydrophobic patches. At initially low angular speeds the liquids do not have the necessary energy to break through these valves, only as the disc's rotational speed is increased do they burst through. Rather elaborate liquid handling has been demonstrated on such centrifugal-force-driven systems [8–10].

Finally, methods utilizing the phenomenon of electroosmosis yield nonpulsating steady flows and typically do not require any moving parts, making them very interesting for a number of reasons. This is explained in more detail in the next section.

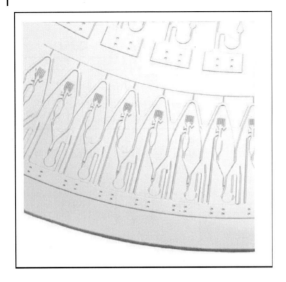

**Fig. 4.7** Detail of a CD containing fluidic structures; hydrophobic patches function as valves (courtesy Gyros AB, Sweden).

4.1.1
**Moving Liquids by Electroosmosis**

At every interface, material characteristics such as charges and forces are not balanced in the way they are inside the bulk, giving rise to several phenomena, including electroosmosis. Let us take a closer look at the interface between a glass capillary or channel and a buffer solution. At the surface of the glass are silanol groups (–Si–OH), which, depending on the pH of the buffer solution, are deprotonated to a greater or lesser extent. Deprotonation results in charge separation, with the negative charges (–Si–O⁻) immobilized on the wall and the protons immediately adjacent to the wall. Most of these positive charge carriers are also immobilized, due to strong electrostatic interactions with the negatively charged wall. However, another, diffuse layer of, again, negative and positive charge carriers forms further into the bulk of the solution. This arrangement of charge carriers, also denoted an electrical double layer, gives raise to an electrical potential between the wall and the bulk of the buffer solution. It is important to remember though, that the entire system is still electrically neutral. A schematic depiction of the double layer model and the resulting potential as a function of the distance from the wall are shown in Fig. 4.8. The potential at the shear plane between the fixed Stern layer and the diffuse Gouy–Chapman layer is called the zeta potential ($\zeta$ potential) [11–13].

The zeta potential is strongly dependent on the chemistry of the two-phase system, i.e., the chemical composition of the wall (material, dynamic or permanent coating, etc.) and the chemical composition of the solution (pH, ionic strength, additives, etc.), as well as on the temperature [13]. If we now apply an external voltage across the buffer solution we can induce a bulk flow of the entire buffer. As an elec-

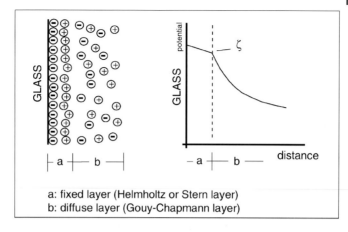

**Fig. 4.8** Schematic of the electrical double-layer model and the electrical potential associated with it.

tric field is applied, the two types of charge carriers are dragged by electrophoretic forces towards the respective electrodes. However, since the negative charges from the deprotonated silanol groups are immobilized at the wall there is a surplus of mobile positive charges in the solution (while the entire system is always electrically neutral). These charges (mainly hydrated protons) drag the entire liquid column towards the cathode. The resulting flow is called electroosmotic (or sometimes electroendoosmotic) flow. We can define an electroosmotic mobility, $\mu_{eo}$, as follows:

$$\mu_{eo} = \frac{\zeta_0 \varepsilon}{4\pi\eta} \tag{4.1}$$

with $\zeta_0$: zeta potential; $\varepsilon$: dielectric constant; and $\eta$: viscosity. Once an electric field, $E$, is applied, this results in the following electroosmotic velocity, $v_{eo}$:

$$v_{eo} = \mu_{eo} E \tag{4.2}$$

Typical field strengths used lie in the hundreds of volts per centimeter, resulting in linear flow velocities on the order of 0.1 to a few mm s$^{-1}$.

To summarize at this point: it is obviously possible to create a flow (induce pumping) by filling a microchannel with a buffer solution and applying a suitable voltage at the channel ends. This relatively easy way is clearly an advantage of electroosmotic pumping, especially compared to the effort it takes to fabricate and run some of the microfabricated pumps presented in section 4.1. Another great advantage is the resulting flow profile. Contrary to hydrodynamic flows, where one finds a parabolic distribution of the flow velocities with the largest velocity at the center of the channel and zero velocity at the walls, electroosmotic flow is generated close to the wall and therefore produces a plug-like profile with a very uniform velocity distribution across the entire cross section of the channel. The proximity to the wall can be assessed by calculating the thickness of the double layer, the so-called Debye length, $\lambda_D$:

$$\lambda_D = \sqrt{\frac{\varepsilon RT}{2F^2 z^2 c}} \qquad (4.3)$$

where $R$: universal gas constant; $T$: absolute temperature; $F$: Faraday constant; $z$: charge number; and $c$: concentration. For a standard borate buffer ($\varepsilon = 78.3\ \varepsilon_0$; $c = 100$ mM, $z = 1$) at standard conditions, it is straightforward to show that $\lambda_D$ is on the order of 1 nm, i.e., more than a thousand times smaller than the channel width or depth. Therefore, virtually the entire cross section has a uniform velocity. Experiments have revealed parabolic and flat flow profiles by imaging a narrow band of released caged dye within a capillary under hydrodynamic and electroosmotic flow conditions, respectively (Fig. 4.9) [14].

The main disadvantage of electroosmotic flow is its strong dependence on the chemistry of the system, which is of course a consequence of the strong dependency of the zeta potential on the chemical state of the system. This dependency makes electroosomotic flow hard to control: every change in pH, dielectric constant, concentration, etc. due to reactions or mixing processes has an immediate effect on the magnitude of the electroosmotic flow. Using buffers can alleviate this situation to some extent and allows for reproducible, constant flow velocities, but only within certain limits. On the other hand, this strong dependency also means that there are many parameters that can be exploited to control and manipulate electroosmotic flow. For instance, lowering the pH reduces the magnitude of the electroosmotic flow and can even suppress it totally. The same can be achieved with a range of chemicals, which can be used to dynamically or permanently coat the channels walls and thereby alter their chemical behavior [15–17]. Here, it is even possible to reverse electroosmotic flow and induce bulk flow towards the anode. Suppression and reversal of the electroosmotic flow are valuable handles for tweaking the performance of electrophoretic separations (see also section 10.4). Finally, electroosmotic flow can be influenced by an external electric field applied through very thin insulating channel walls [18].

Apart from the direct pumping by electroosmosis, this phenomenon can also be used in an indirect way [19–23]. Only two examples of such implementations are briefly described here. Let us look at a T-type channel intersection (Fig. 4.10). If we apply a voltage from reservoir 1 to reservoir 2 we find that all the liquid follows this path, i.e., the fluid flow follows the current flow. If we now chemically modify the surface in the side-arm labeled C2, so that electroosmotic flow in this section is suppressed, but still have the current flowing as before, the following happens: at the intersection the fluidic flow generated up to this point wants to follow the current flow into the side-arm. However, this is now almost impossible because electroosmotic flow is chemically suppressed in this part. Consequently, pressure builds up at the intersection and the fluid follows the path of least resistance, i.e., becomes divided into a part flowing towards C2 and a part flowing into the field-free region towards C3, the ratio depending on the ratio of the actual

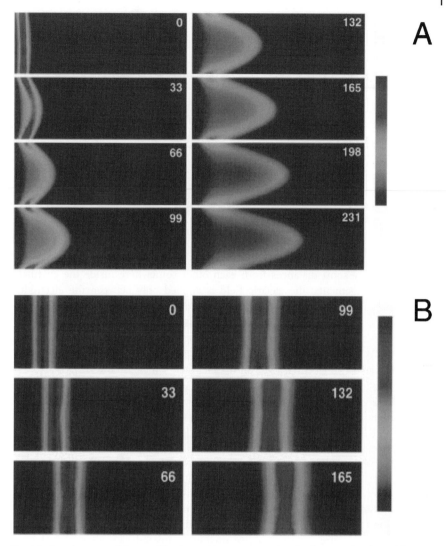

**Fig. 4.9** Development of flow profiles imaged by using caged and subsequently released fluorescent dyes. A) hydrodynamic flow; B) electroosmotic flow (reprinted with permission from [14]. Copyright (1998) American Chemical Society).

flow resistances in these two parts. In this fashion, we have induced hydrodynamic pumping (with a parabolic flow profile) by means of electroosmotic flow. We have thus utilized the relative simplicity of generating electroosmotic flow to facilitate liquid movement in a different, field-free part of the channel network. In many applications it is preferable to avoid exposure to high electric fields, e.g., when working with living cells. Electrodes used for applying the voltage do not necessarily need to be placed at the ends of the channels, but can be placed at al-

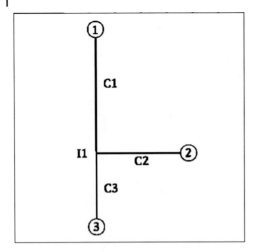

**Fig. 4.10** Channel layout for electroosmotically-induced hydrodynamic pumping. C2 is chemically surface-treated to suppress electroosmotic flow (reprinted with permission from [19]. Copyright (1998) American Chemical Society).

most arbitrarily short distances from each other somewhere inside the channels. In these arrangements, the electric field is applied only between the electrodes, and the rest of the channel is field-free. Electroosmotic flow is also generated only in the space between the electrodes, but, because the liquid has to move somewhere, this induces hydrodynamic flow outside the region between the two electrodes [22]. A variant of this setup uses porous packings or porous polymer immobilized between two electrodes within the channel to increase the surface area and thereby the magnitude of the electroosmotic flow [24–26]. However, whenever electrodes are placed inside channels, care has to be taken to allow gas bubbles, which might be generated by electrolysis at the electrodes, to leave the system, because they can grow and interrupt current and fluidic flow.

## 4.1.2
### Mixers

The importance of mixing in chemical microsystems is discussed in chapter 3.3. The goal is to arrive at a homogeneous blend of two solutions in as little time as possible. We also stressed that passive mixing on microdevices depends solely on diffusion as the transport mechanism. All strategies to improve mixing have therefore looked into the possibilities of either reducing the diffusion distances or inducing some kind of extra movement (typically lateral to the flow direction). Alternatively, active mixers, in which external energy is used to induce turbulence, have been investigated as well. Examples of how mixers are implemented on microsystems are given in this section.

The Einstein–Smoluchowski equation describes the dependence between the diffusion distance and the diffusion time. If we reduce the diffusion distance by a factor of 2, the diffusion time is reduced by a factor of 4; if the distance is reduced 10-fold, the time is reduced 100-fold. To obtain short diffusion distances we could design narrow or high-aspect ratio channels to begin with. However, just narrowing chan-

**Fig. 4.11** Principle of a laminating mixer – the main idea is to minimize the necessary diffusion lengths.

**Fig. 4.12** Experiment and simulation of a laminating mixing structure (courtesy U. D. Larsen, Chempaq [36]).

nels results in higher fluidic resistance and may therefore not be a good choice. Channels of high aspect ratio (narrow and deep), on the other hand, are harder to fabricate and require special machining technology. One possible strategy is to work with standard-sized channels, split each channel into an *array* of smaller channels (thereby circumventing the high fluidic resistance of just *one* small channel), and then merge them again in such a way that the split flows of solution A get interlaced with the split flows of solution B (Fig. 4.11). Although it looks simple enough in this schematic picture, it is actually quite difficult to implement, because it requires adding a third dimension to the typically planar, quasi two-dimensional, microliquid device. One of the two types of solutions has to be brought in from above or below to properly interlace with the other solution. Fig. 4.12 shows a microphotograph of a mixer implemented in such a fashion, together with results from a simulation of the same device. An indicator solution (dark) is brought in from below through an arrangement of openings to be interlaced with an acid solution (clear), resulting in a grey color. From the dark-colored streaks we can see that it takes some time for diffusion to mix both solutions completely. This process takes much longer at the walls, because there the indicator is only touched on one side by the acid. The simulations nicely match the experimental results. One of the main disadvantages of such a mixer design is its large volume. If two small well-defined plugs of chemicals are to be mixed in such a mixer, the excess volume might lead to strong dispersion

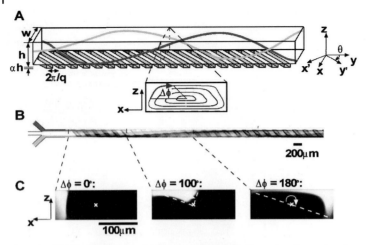

**Fig. 4.13** Fluidic streamlines obtained with obliquely grooved surfaces: A. schematic streamlines; B. experiment (top-view); C: cross-sectional views with one fluorescing fluid (reprinted with permission from [27]. Copyright (2002) American Association for the Advancement of Science).

and hence to dilution of the combined plug after mixing, which in turn could slow down or prohibit a reaction or result in a low signal-to-noise ratio for detection.

As mentioned in Chapter 3, the Peclét number (*Pe*) is a good indicator of whether diffusive mixing alone is sufficient, given a certain channel width (i.e., a certain minimum diffusion length) and a certain linear flow velocity. The higher the Peclét number the harder it is to achieve mixing by diffusion, especially within a limited amount of channel length. In fact, the necessary channel length increases linearly with the Peclét number [27]. For this reason, ways to induce a lateral transport component (i.e., an additional flow vector more or less orthogonal to the main flow direction) to complement diffusion and improve mixing are necessary. One such possibility is to machine grooves into the channel bottom, where these grooves are of certain depths, arranged at a certain angle with respect to the main flow direction, and arranged in a number of patterns [27, 28]. For pressure-driven flow, these grooves introduce anisotropic flow resistances into an otherwise isotropic system. Fluids experience less resistance when flowing along the ridges and valleys constituting the grooves than when flowing perpendicular to them. In this way, a rotational element is introduced, resulting in corkscrew-like fluidic streamlines (Fig. 4.13). A careful arrangement of different types of grooves (mainly differing with respect to their orientation) can produce chaotic stirring, which allows mixing in much less time and shorter channel length [27]. Similar ideas were also tested with electroosmotic flow. Here, due to the oblique arrangement of the grooves there is a tiny electrical field drop along the grooves, inducing an electroosmotic flow component that is lateral to the main flow direction. Experiments showed that it was again not enough to use only one kind of groove. Instead, alternating grooves with different orientations greatly improved the mixing efficiency (Fig. 4.14) [28].

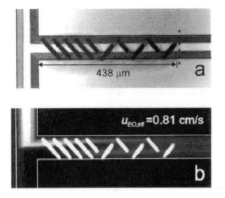 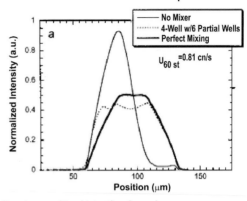

**Fig. 4.14** Electroosmotically-driven mixer with grooved wells micromachined into the channel surface (reprinted with permission from [28]. Copyright (2002) American Chemical Society).

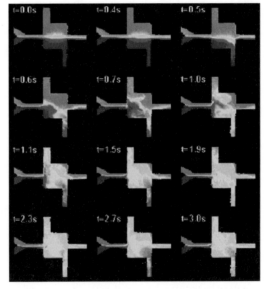

**Fig. 4.15** Sequence of mixing using electrokinetically induced flow instabilities; from left to right are shown the Y-junction inlet, the main mixing chamber with two channels connecting to the AC electrodes, and the common exit channel (reprinted with permission from [29]. Copyright (2001) American Chemical Society).

All the above examples are passive mixers, relying solely on energy from within the system to facilitate mixing. Active mixers can also be designed, in which external energy in some form induces chaotic behavior, leading to improved, accelerated mixing. One example is shown in Fig. 4.15, where a sinusoidally oscillating electric field is used to induce flow instabilities, which in turn lead to stretching and folding of the fluids and hence mixing. In the design used, a mixing chamber has two electrodes connected to it, where the oscillating electric field can be applied. Given the right set of parameters, mainly with respect to applied field strength and oscillating frequency, flow instabilities can be observed manifesting themselves as seemingly random and transverse velocity fluctuations. Again, these additional velocity components lead to chaotic behavior, resulting in enhanced mixing within a shorter timeframe [29].

## 4.2
## Injecting, Dosing, and Metering

A very important fluidic handling function is the ability to dispense very defined (small) amounts or volumes of solutions, repeatedly, on demand, and very reproducibly. 'Defined' in this context can refer either to a volume whose size is not exactly known but which is always the same or to a volume whose size is exactly known and given by, e.g., a particular geometric arrangement of channel segments. The latter is also known as 'metering'. For example, in the microfluidic structure in Fig. 4.7, liquid was introduced into the channels through an opening hole, and all available channels were filled by capillary action. Fluid flow was stopped by burst valves based on hydrophobic patches. After filling, the CD was set into motion, inducing flow where it was not prohibited by the hydrophobic patches. In this way, fluid was removed from excess channels leaving a volume of fluid behind, as defined by the channel geometry between the excess channel and the hydrophobic valves. Thus, these volumes were metered off. Any similar concerted action of pumping and valving involving geometrically defined volumes can in principle be used to meter off defined amounts of solutions.

For many analytical techniques it is beneficial to be able to inject small, well-defined amounts of sample solutions (see also chapter 10). This is also referred to as aliquoting. Popular implementations of an aliquoting or injection function on microchips make use of a channel cross or a double T arrangement along with electrokinetic fluid manipulation to achieve this. Let's have a closer look at what is going on at a simple channel cross (Fig. 4.16, left). A popular approach to describing the electrical behavior of a fluidic channel network is to use electrical equivalent circuits, e.g., representing the channel cross by several nodes and resistors (Fig. 4.16, right). There are five nodes – four at the terminal ends and one at the intersection. Connecting the nodes are four resistors representing the impedance in the four channel segments. The only external control we have is the voltage we apply at the four terminal nodes, whereas we can only calculate the voltage at the inner node, once we have measured and established the individual impedances. Now, with a few relatively simple relations, such as Ohm's Law and the Kirchhoff rules, we can calculate which voltages need to be applied in order to have current following a specific pathway through the channel network. And, assuming similitude between the electric field and the fluidic flow, transport of bulk liquid and chemical species should also follow this pattern.

**Chip Layout**                                                **Circuit Representation**

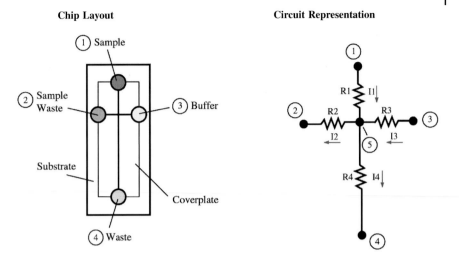

**Fig. 4.16** Schematic chip layout with a simple cross for injections (left) and the equivalent electrical circuit representation (right).

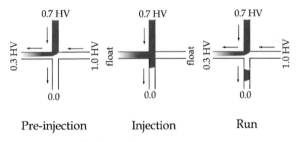

**Fig. 4.17** Principle of gated injection.

Fig. 4.17 illustrates the application of these ideas to the implementation of a so-called gated (electrokinetic) valve or injector [30, 31]. The main waste reservoir at the end of the main channel is set to ground. Thus, it functions as the anode to which the electroosmotic flow is directed. All other reservoirs are at higher positive potentials. The topmost reservoir contains the sample and, to ensure pumping out of the reservoir, the potential is set to a somewhat large value, indicated here as a percent fraction of a specific limit value. For the same reason, the sample waste reservoir is set to a relatively low voltage to receive the sample flow. The remaining reservoir is filled with buffer and is set at a slightly higher voltage than the sample reservoir. This allows flow out of this reservoir, to keep sample from entering (bleeding into) the main channel by flowing right across the junction and into the sample waste reservoir. At the same time, this reservoir also provides flow of fresh buffer into the main channel. The distribution of voltages described so far corresponds to the situation before an injection. Depending on the specifications of the power supplies used, the simplest way to facilitate injection is to remove the power supplies at the sample waste and the buffer reservoir from the electric circuit. This brings their potential to exactly the same potential as at the intersection at this moment, thereby preventing any flow to or from the side arms. Voltage is still applied at the other reservoirs, allowing sample to flow into the main channel. The injection step is finished as soon as the original voltage distribution is established again, thereby cutting off the sample flow into the main channel and releasing a plug of sample flowing towards the anode. Because the injected volume is mainly determined by the timing of the injection sequence and sample can enter the main channel for as long as this electrokinetic valve is open, this injection procedure is called 'gated injection'. There is one caveat, however: because electrokinetic forces (i. e., electroosmosis and electrophoresis) are used to inject the sample, the injection is biased, meaning that during the time of injection, positively charged ions are injected to a larger extent than neutral species, and neutral species to a larger extent than negatively charged ions. This is, of course, because the cations have an electrophoretic velocity in the same direction as the electroosmotic flow, while neutrals have no electrophoretic velocity, and anions have an electrophoretic velocity opposed to the electroosmotic flow. This fact needs be taken into consideration when designing a method for quantitative analysis involving one or more gated injection steps.

There is an alternative to the gated injection method, which also utilizes the cross layout already mentioned. Here, the sample reservoir is on one of the side-arms, and prior to injection, sample is continuously pumped across the intersection to the waste reservoir (Fig. 4.18). Additionally, the voltages are arranged so that there is also flow from the buffer reservoir and the reservoir at the end of the main channel towards the intersection and on to the sample waste reservoir. This is again to avoid premature bleeding of sample into the main channel. By adjusting the voltages (i.e., the flow rates) one can pinch the sample stream more or less on its passage through the intersection (see also Fig. 3.6 on flow focusing). Therefore, this technique is often called 'pinched injection' [30, 31]. The main differences with respect to gated injection are:

- The injection volume is pre-defined and fixed. Only the volume defined by the junction geometry and the pinching is injected once the voltages are switched accordingly (Fig. 4.18).

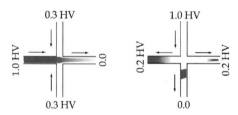

Load position          Inject & Run position          **Fig. 4.18**  Principle of pinched injection.

- If the loading step is given sufficient time, then the composition of the sample solution present at the intersection is the same as the composition of the original sample solution, and consequently there is no electrokinetic bias upon injection.

A variation of the pinched injection is the so-called double-T injection, where the two side-arms of the cross geometry are offset by a certain distance, thereby increasing the volume to be injected. Although this volume is larger than in the original pinched injection, it is again fixed and cannot be changed once the chip is fabricated. (Of course, various degrees of pinching slightly alter the injected volume). Other variants have also been investigated, in which either geometries (channel dimensions) or voltage switching protocols have been altered [32, 33]. These injection procedures are all based on electrokinetic flow control and it is relatively cumbersome to implement similar schemes with pressure-driven flows. However, it is also possible to design a gated injection valve with a field-free injection region using electroosmotically induced hydrodynamic flow.

Still, dosing is of course also possible with other means than electrokinetic fluid handling. Often, dosing of a defined volume can be achieved with pumps or pump-like microstructures, in which a single stroke of, e.g., a membrane, displaces a certain amount of fluid out of a pump chamber, through a valve or just through some orifice. Well-known implementations of such micromechanical dosing systems are ink-jet printer cartridges. In these cartridges the actuation can be achieved by a gas bubble, which is initiated on top of a heating element, causing displacement of fluid through an opening as the bubble grows. When the current is switched off the bubble shrinks again, allowing fresh material to enter the reservoir for the next dosing action. Alternatively, piezoelectric actuators are often used, actuating a membrane to push on an enclosed volume of fluid in order to press part of it out of an opening. Piezoactuators can be operated at high frequencies, allowing high droplet generation rates. An example of a piezoelectrically operated microdispenser is shown in Fig. 4.19 [34]. Droplets generated from such systems have also been used in chemical settings, e.g., in connection with acoustically levitated droplets, which constitute wall-less chemical containers and which can be loaded with tiny droplets from piezodispensers to allow many chemical reactions to take place within the large levitated droplet [35].

In general, dosing with microsystems is considered to be of great importance in the development of medical microdevices, which not only diagnose and monitor certain body functions, but, at the same time, also can administer drugs in precise amounts to regulate those functions. Such systems will allow patients to enjoy more independence from stationary medical treatment and probably also reduce the overall amount of drugs necessary for a treatment.

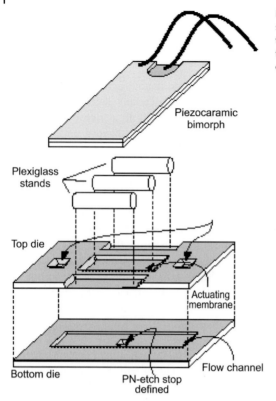

**Fig. 4.19** Schematic of a dispensing microsystem using piezoactuation (reprinted with permisssion from [34]. Copyright (1999) Institute of Physics).

Piezocaramic
bimorph

Plexiglass
stands

Top die

Actuating
membrane

Bottom die

PN-etch stop
defined

Flow channel

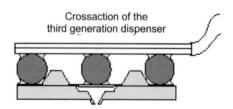

Crossaction of the
third generation dispenser

## 4.3
**Temperature Measurement in Microfluidic Systems**

Apart from precise dosing of fluids in a microsystems, it is often desirable and sometimes necessary to measure the temperature. Microsystems can be used for production or analysis of biological or chemical components. In both cases, the temperature dependence of chemical and biological reactions makes temperature measurement desirable or even necessary.

When a microfluidic system is used as an analyzer, the system is concerned with measuring the concentration of biological or chemical species; this measurement is almost always directly or indirectly temperature dependent. Most often,

**Fig. 4.20** A microfluidic system that is used as an analyzer. Similar to the reaction in Eq. 4.4, the input reagents are a sample A and an additional reagent B, the product AB of the reaction is the waste output of the system to the right.

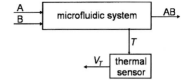

measurement of concentration relies on a biological or chemical reaction that occurs within the microsystem. Such a reaction can be endothermic (taking up heat from the surroundings) or exothermic (releasing heat into the surroundings), and the reaction speed is also temperature dependent. The reaction can typically be described as

$$A + B \xrightarrow{k} AB + Q \, , \tag{4.4}$$

where A and B are chemical reactants, AB the reaction product, $k$ the reaction constant, and $Q$ the generated heat. Reaction (4.4) described here is exothermic, because $Q$ was chosen to be positive.

An important observation is that the rate constant $k$ is temperature dependent. This means that if the temperature changes by any means, whether by external influences or by the reaction heat $Q$ itself, the measured results also change. At the wrong temperature, the reaction might occur too late in the microsystem, so that the reagents get flushed out of the system, or the reaction might occur too early. Either way, the target species concentration is not measured accurately.

To monitor the reaction, the use of a temperature sensor is advisable (Fig. 4.20). Measuring the temperature is also the first step needed for active control of the reaction or for temperature compensation, if control or compensation is needed.

### 4.3.1
### Microreactors

Chemical and biochemical microsystems can also be used for production of components; here, the system is called a microreactor. Temperature sensors play an important role in such microreactors, because chemical reactions are usually strongly temperature dependent, as discussed above. A relatively simple microreactor is sketched in Fig. 4.21.

The reactor has incoming reagents A and B, and product AB leaves the reactor. The temperature $T$ in the microreactor can be monitored with the thermal sensor, which outputs a signal $V_T$ that corresponds to $T$. The microreactor is also equipped with a heater/cooler unit to make it possible to keep the reactor at a specified temperature. Together with the thermal sensor, the heater/cooler unit forms a closed-loop controller for the microreactor. Such a controller can be very useful, for instance for polymerase chain reaction (PCR), where a microreaction chamber cycles through several different temperatures in sequence.

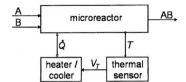

**Fig. 4.21** A microreactor with two input reagents A and B, and output product AB. A heater/cooler unit removes heat $Q$ from the microreactor. A thermal sensor measures the reactor temperature $T$ and supplies this information to the heater/cooler unit via its output signal $V_T$.

If the temperature change is small and negligible for the reaction under consideration, then the thermal sensor is not needed. If however the temperature change is considerable, then the change must be recorded for further processing or used for direct temperature compensation. Finally, the temperature change itself can be the signal of interest, when the reaction heat is measured to find out how much of the product was produced. In this case, it is important that no extra heat enters or leaves the microreactor, so that the rise in temperature is proportional to the internally produced heat. If a microsystem is used this way, it is called a microcalorimeter.

Apart from the interest in measuring temperature as a prime goal, temperature measurement is also important in other areas, especially for flow measurement.

### 4.3.2
### Temperature Sensors for Microsystems

Of the many different types of temperature measuring principles, only a few are of principal interest for chemical and biochemical microsystems. These main sensing principles are measuring a change in resistance of a material, measuring a thermoelectric voltage, and exploiting the thermal dependence of electronic parameters of *p-n*-junctions. Which sensor is used in a given application depends in general on the needed sensitivity, resolution, range, bandwidth, signal-to-noise ratio, cost, and last but not least, the size of the sensor. Usually a sensor for a chemical or biochemical microsystem is as small as possible. Small size has the added advantage of a small heat capacity and thus a relatively fast response of the sensor. Finally, it is desirable to be able to integrate the fabrication of the sensor into the production process of the microsystem. Sensitivity, resolution, range, signal-to-noise ratio, and cost depend on the individually chosen sensors, which are discussed below.

### 4.3.3
### Resistance Temperature Detectors

#### 4.3.3.1 **Metals**
If the change in the resistance of a metal with temperature is measured, the resulting sensor is known as a resistance temperature detector (RTD). Platinum is the metal of choice, because it is corrosion resistant and stable over a wide temperature range. Platinum resistance sensors are the industry standard, due to their accuracy and their reproducibility [37]. Within a temperature range of 15 to

725 K the accuracy of platinum resistance sensors can be as low as 0.01 K [1]. Resistance sensors can be integrated into a microfluidic sensor with comparative ease by depositing thin films. Not only platinum is used for microsystems but also other metals can be deposited, for example, a nickel–iron alloy [38].

Platinum resistance sensors are ubiquitous in microfluidic systems for polymerase chain reaction (PCR), where a certain temperature–time profile has to be driven in a miniature vessel. The resistance sensor is often deposited between integrated heaters, so that the sensor together with the heaters form part of a closed-loop heat control. A typical example of such a PCR chamber with temperature controller based on platinum resistors is the microsystem shown in Fig. 4.22 and Fig. 4.23.

An example of a platinum resistance sensor is one used by Zhan *et al.* [39]. They present a microfluidic system for PCR, in which both the heaters and the sensor were made from a deposited and then patterned thin film of platinum. The temperatures used in the microvessel were slightly below 100 °C, and the accuracy achieved in the system was 0.5 K.

#### 4.3.3.2 Nonmetals

Resistance temperature sensors can also be produced from nonmetals. Ceramics and oxides on the one hand and semiconductors on the other hand are especially interesting, for different reasons.

A disadvantage of the platinum resistance sensor is its relatively low sensitivity, less than 1% $K^{-1}$. Some ceramics and oxides offer much greater sensitivity, 4%–6% $K^{-1}$, which makes them attractive as temperature sensors, because larger sig-

**Fig. 4.22** PCR chip. A microsystem with integrated resistive heater and resistance temperature detector (RTD) is shown from above. The RTD is the single central wire, flanked by several heating wires to the left and the right. The wires are deposited platinum. A cross section is shown in Fig. 4.23 (courtesy El-Ali and Wolff, MIC).

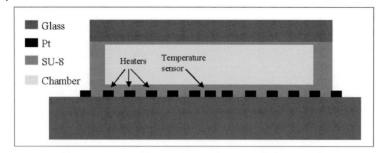

**Fig. 4.23** Chamber cross section. The PCR microsystem of Fig. 4.22 is depicted schematically in cross section (courtesy El-Ali and Wolff, MIC).

nals for the same temperature change mean a higher signal-to-noise ratio. These nonmetal resistance temperature sensors are known as thermistors. There are two major types, one having a negative temperature coefficient (NTC thermistor) and the other having a positive temperature coefficient (PTC thermistor). The drawback of thermistors is their strong nonlinear temperature dependence, which calls for more intensive signal processing. Thermistors have an interchangeable temperature tolerance of 0.1 K, although the accuracy can be increased by individually calibrating the sensors.

Using silicon as a sensor material has the advantage that the sensor can be easily integrated into a silicon microfluidic system. Hauptmann [40] describes a silicon resistance temperature sensor based on the principle of spreading resistance. This sensor measures the resistance of silicon between a flat back contact and a point front contact. The dependence on temperature is slightly nonlinear, although not as drastically as for a thermistor. The temperature coefficient or the sensitivity of the sensor is 0.8% $K^{-1}$, which is lower than that for thermistors, but still higher than that for platinum. Such sensors can be acquired commercially as discrete devices from Philips Semiconductors, Eindhoven, The Netherlands; examples being devices from the KTY84-1 series.

Resistance temperature detectors are usually measured by a four-wire technique. To measure resistance, the minimum number of connections to a resistor is two, one at each terminal. A voltage is then established across the resistor and the flowing current is measured, which allows the resistance to be calculated with Ohm's law. A systematic error, however, is introduced in this measurement setup, which is due to the additional connection cable length. This systematic error can be avoided by using a four-wire measurement technique. Here, there are two connection cables on each terminal. A constant current is sent through one set of cables, while the resulting voltage drop is measured with the other set. The four-wire technique is superior to the two-wire technique, because the connection wire resistance is much lower than the input resistance of a voltage meter, and voltage is measured instead of current.

4.3.4
**Thermocouples**

Thermocouples are very popular temperature sensors, which rely on a thermoelectric phenomenon, the Seebeck effect. The essential parts of a thermocouple are two wires of different metals. If the wires are welded together at one end and a temperature difference is applied between this point and the other two ends, then a small voltage proportional to the temperature difference is generated. This is known as the Seebeck effect. In practice, the other two ends are also welded together, and one of the wires is cut open for voltage measurement. This procedure prevents additional thermovoltages from forming between the thermocouple ends and the voltage measurement leads.

Different metal or metal-alloy combinations are useful for thermocouples. Some combinations, such as chromel–constantan and iron–constantan, optimize the sensitivity of the thermocouple, but others, such as chromel–alumel copper–constantan, and platinum/rhodium–platinum, make it possible for the thermocouple to operate over an especially large range or within an unusually cold or hot temperature region.

In comparison to a resistance temperature sensor, thermocouples have the advantage that they work over a very large temperature range, of up to 3000 K (Kovacs 1998). Thermocouples are especially suitable when a low mass is needed and temperature differences are to be measured. The various thermocouple material combinations result in different values of sensitivity and accuracy. The platinum/rhodium–platinum thermocouple, for instance, has a sensitivity of only 6 μV K$^{-1}$ with a relative accuracy of 0.25%. The chromel–alumel thermocouple has a tenfold sensitivity of 60 μV K$^{-1}$, also accompanied by a higher relative accuracy of 1% (van Putten 1996). The absolute accuracy of a calibrated thermocouple is in the range of 0.5 to 2 K.

Thermocouples can also be produced from moderately doped semiconductor material (Kovacs 1998), which makes it comparatively easy to integrate them in a microfluidic chip.

4.3.5
**Semiconductor Junction Sensors**

Silicon, which has long been the most abundant material used for the micromachining of microfluidic devices, can be used as a temperature sensor in more than one way. A silicon resistance sensor was mentioned above; here, exploitation of the temperature dependence of one or more *p–n* junctions is discussed. Several electronic variables of a *p–n* junction depend on temperature; the saturation current of a diode and the base-emitter voltage of a transistor are examples.

A diode can be used as a temperature sensor in a similar fashion as a resistance temperature sensor. A certain constant current is sent through the diode, and the voltage across is measured with a voltmeter (Kovacs 1998, Yeager and Courts 2001). The sensitivity of such a temperature-measuring diode is 2.0–

2.4 mV K$^{-1}$ (Kovacs 1998). The range of the diode sensor is from 1–500 K, and diode sensors can achieve an accuracy of 0.1 K, when calibrated.

Using the temperature dependence of the base-emitter voltage of a transistor makes it possible to design a compensation circuit that produces a sensor signal that is proportional to the absolute temperature. The base-emitter voltage ($V_{BE}$) is given by

$$V_{BE} = \frac{kT}{q} \ln\left(\frac{I_C}{I_S}\right) \tag{4.5}$$

where $k$ is Boltzmann's constant, $T$ is the temperature, $q$ is the elementary charge, $I_C$ is the collector current, and $I_S$ the saturation current. The base-emitter voltage itself depends on temperature in a rather complicated way, mostly because the saturation current ($I_s$) depends on several temperature-dependent variables, such as the intrinsic carrier density, the diffusion coefficients, and the diffusion length. However, by looking at the difference between the base-emitter voltages from two transistors having different collector currents, the temperature dependence of the saturation current can be eliminated. The voltage difference is found to be

$$\Delta V = V_{BE1} - B_{BE2} = \frac{kT}{q} \ln\left(\frac{I_{C1}}{I_{C2}}\frac{I_{S2}}{I_{S1}}\right) \tag{4.6}$$

which leaves the resulting circuit output signal ($\Delta V$) proportional to the absolute temperature. Such a circuit is therefore often abbreviated PTAT.

Such a PTAT circuit can have an accuracy of 1 K at a sensitivity of 10 mV K$^{-1}$ within a range of –25 °C to 85 °C (about 250 to 360 K). Discrete versions of this circuit are also available commercially, for instance, as part number AD549 from Analog Devices, Norwood, MA, USA. Frank [41] describes how PTAT temperature sensors are embedded into a larger integrated circuit to provide means for detecting whether temperature, voltage, or current are out of limits. The overall product is a so-called smart power integrated circuit. It is also possible to integrate a diode or a PTAT circuit into a microfluidic silicon system.

### 4.3.6
### Temperature Sensors Built on Other Principles

In addition to the widespread and well-known temperature sensors mentioned so far, devises based on other principles are also used to measure temperature in microfluidic systems.

Fluorescence thermometry [42, 43] and liquid-crystal thermometry allow spatially resolved fluid measurements. Both methods are based on the distribution of temperature-sensitive elements all over the microfluidic system. When liquid crystals are used as elements, different colors of the crystals indicate different temperatures. When fluorescence is used, the typical fluorescent decay time indicates the temperature. Chaudhari *et al.* [44] reported temperature measurements in a

PCR vessel using liquid crystals. They achieved a measurement accuracy of 0.1 K within two rather small temperature intervals of 54.5–55.5 °C and 94–95 °C.

Acoustic waves in solids can also be used for temperature sensing [45]. Surface acoustic waves and plate waves are mainly used, where the waves travel in a material interfacing the fluid in the microsystem. The waves can be detected electronically, because the material the waves travel in is piezoelectric. The output signal of the devices is a frequency, which makes it convenient for electronic signal processing. The frequencies are in the megahertz range. Although the sensitivity achieved is relatively low at a maximum value of about 200 ppm K$^{-1}$, the accuracy is relatively high at about 0.2 K.

Temperatures can also be measured mechanically by using a bimetallic strip. Such a strip consists of two materials with different thermal expansion coefficients, so that the strip bends when the temperature changes. A bimetallic strip is usually used as a thermally activated switch (Kovacs 1998). However, the strip can also be used to measure temperature continuously by measuring the extent of bending with a strain gauge. Marie et al. [46] used micromachined cantilevers for temperature control in a PCR chamber, with the advantage of having the temperature sensor directly immersed in the fluid. The second advantage is the convenience of being able to use another cantilever in addition to the ones already in use for a different purpose.

### 4.3.7
### Conclusion

If it is necessary to measure temperature in a microfluidic system, then the choice of the thermal sensor should be based mainly on whether it is a semiconductor-based microsystem, in what range the temperature has to be measured, and to what accuracy, while keeping cost in mind.

The temperature sensors most often used in microfluidic devices are platinum resistance temperature detectors and thermocouples, although other sensors, such as thermistors and diodes, might be increasingly used for mass-produced devices.

Depending on the microsystem under consideration, it might be convenient to choose a sensor principle that is easily integrated into the system, such as a silicon diode in a silicon microsystem or a bimetallic strip in a cantilever-based measurement system.

### 4.4
### Optical Sensors

Optical sensors also play an important role in microsystems. We all use visual inspection of liquids in our everyday life, e.g., estimating the strength of coffee from its color. Coffee appears dark, because light is absorbed by the liquid, so a very dark color corresponds to a high concentration of absorbing molecules. This is one of many principles (absorbance detection) that are used for optical detection in chemical analysis systems. A wide range of other optical phenomena are used to relate a concentration to an optical signal. Among these, luminescence is important: a molecule is excited to a higher energy state and relaxes to the ground state by emission of optical energy. The optical power of the emission can be related to the concentration of molecules and hence can be used for detection. The various optical measurement schemes are discussed in sections 4.4.2, 4.4.3, and 4.4.4, after a short introduction to the instrumentation that is necessary in a complete chemical analysis system.

### 4.4.1
### Instrumentation

In most microanalytical systems neither the light source nor the photodetector is integrated into the same substrate as the fluidic channel network. The reason is that monolithic integration of all components necessary for performing a total chemical analysis is very complex. Thus, most research groups have focused on a few aspects of a whole system, such as the fluidics, in which a typical system consists of an assembly of different modules for which efficient interconnections are very important. The most widely used optical interconnection methods are optical fibers and free-space optical elements such as lenses, mirrors, and filters to guide light into the channel network and also to collect optical signals from the same fluidic channels. Typical modules are outlined in Fig. 4.24. Examples of devices are given in the following sections, where optical elements such as planar waveguides and photodiodes are integrated with a microfluidic channel network.

A light source is often needed (not necessary for chemiluminescence measurements), and a very popular choice is the laser (light amplification by stimulated emission of radiation), because it has a high optical power and is monochromatic, which means that the light is confined to a very narrow wavelength range. A major limitation

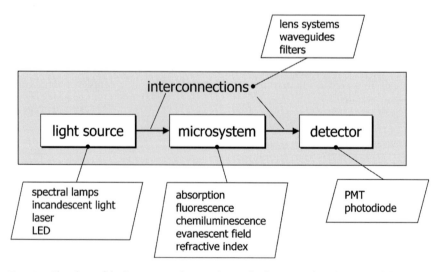

**Fig. 4.24** Flowchart of the basic units of a typical microfluidic system based on optical detection.

of lasers is that they are often large and expensive. This is especially true for lasers operating in the ultraviolet range (UV), which is about 200–400 nm. In this range other light sources such as spectral lamps and incandescent lamps are used. Another popular choice is the light-emitting diode (LED), because of its very small size (a few millimeters on each side). Its smaller optical output power is often a limitation.

On the detector side the two most popular types are the photomultiplier tube (PMT) and the semiconductor photodiode. The photomultiplier tube internally amplifies the detected signal, which is the reason for the name 'photomultiplier': the incoming signal is optically 'multiplied', which enables measurements at low light levels.

## 4.4.2
### Absorption Detection

Absorption of light is one of the most fundamental aspects used in optical detection and is thus very important to be familiar with. Absorption means that optical energy (photons) is converted into another type of energy, e.g., into thermal energy in the absorbing solution – which results in an increase in the temperature of the solution. In the ultraviolet and visible wavelength range (180–700 nm) light absorption is typically due to an electronic transition to a higher energy state (Fig. 4.25). (In fluorescence, section 4.4.4, a fraction of the optical energy is not converted into heat but is re-emitted as photons having lower energy and hence longer wavelength.)

Fig. 4.25 (left) shows a schematic drawing of an absorption measurement. An absorbing solution with a concentration $c$ is contained in a transparent cuvette, and light of a fixed wavelength is passed through it. The so-called absorbance value $A$ can be calculated from the transmittance, $T(c)$, which is the ratio between the optical power with light attenuation due to absorption by the solution and without attenuation due to absorption. This value (which is of course smaller than 1) can be estimated from the ratio between $P(c)$ and $P_0$, as seen in Fig. 4.25 (left). This is, however, a poor estimate, because contributions in addition to absorption by the so-

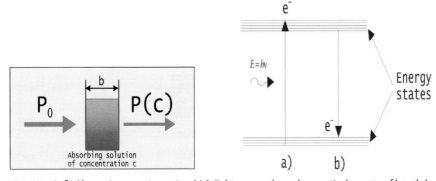

**Fig. 4.25** Left: Absorption experiment in which light passes through an optical cuvette of length $b$. Right: Absorption process by which optical energy ($E = h\nu$) is converted into thermal energy. a) Radiative excitation. b) Radiation-less decay.

lution also decrease the transmitted optical power. For example, reflections at interfaces of different refractive indices reduce the transmitted optical power. To avoid this problem, a reference measurement is generally made with a transparent solution in the cuvette $P_{\text{solvent}}$. This is typically the solvent that was used to prepare the absorbing solution. Hence, the transmittance is given by

$$T(c) = \frac{P(c)}{P_{\text{solvent}}} > \frac{P(c)}{P_0} \tag{4.7}$$

The absorbance value $A$ is calculated with the relation:

$$A(c) = -\log[T(c)] = \log\left[\frac{P_{\text{solvent}}}{P(c)}\right] \tag{4.8}$$

and the absorbance is seen to increase with a decrease in the transmitted optical power. A very important formula is the Lambert–Beer law, which states that the calculated absorbance is proportional to the concentration of the absorbing solution and to the optical path length $b$, used during the measurements.

$$A(c) = \varepsilon bc \tag{4.9}$$

$\varepsilon$ is the molar absorptivity. This value is a property of the chemical species and also depends on the wavelength. It is thus necessary to choose a wavelength with a high value of $\varepsilon$ to obtain the greatest signal for a given concentration and optical path length. The Lambert–Beer law is generally valid for low concentrations and monochromatic light.

Eq. 4.9 indicates that an increase in the optical path length is desirable, because it leads to an increase in the calculated absorbance value and allows measurement of lower concentrations $c$. The sensitivity of absorbance measurements thus scales poorly with miniaturization of the channel dimensions, because it often results in a decrease in the optical path length. Fig. 4.26 shows a possible way of overcoming problems with a short light path length.

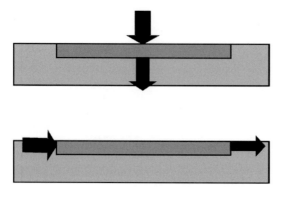

**Fig. 4.26** Top: Absorbance measurement perpendicular to the plane of the microdevice, resulting in a short path length. Bottom: Absorbance measurement in the plane of the microdevice, resulting in an increased optical path length.

**Fig. 4.27** Photograph of a capillary electrophoresis device with integrated planar waveguides for absorption detection. The inset picture shows a 750-μm-long absorption cell [47].

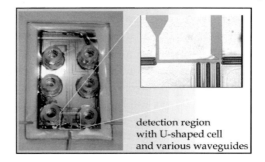

detection region
with U-shaped cell
and various waveguides

If the measurement can be done in the plane of the device, a longer path length is possible, because it is not restricted to the depth of the channel, which is ultimately limited by the substrate thickness. In a practical application the channel depth is typically less than 100 μm, because a low volume is desired. This principle was used in a device in which optical waveguides were integrated with microfluidic channels to guide the light in the plane of the chip (Fig. 4.27). Planar waveguides are transparent structures that consist of a material of high refractive index surrounded by a material with a lower refractive index. Guidance of light is thus obtained by total internal reflection, which is also used in optical fibers.

A photograph of the device (Fig. 4.27, left) shows the six reservoirs that contain the chemicals during analysis at the top of the device. Electroosmotic pumping and separation by means of capillary electrophoresis can be performed by switching the voltage between electrodes that are placed in the reservoirs (see section 4.1). The inset picture on the right shows a section of the channels that is interrogated by two planar waveguides. This configuration increases the optical path length by enabling absorbance to be measured along the channel. The absorption path length is 750 μm, which is more than 50 times the channel depth (13 μm). The left picture also shows how light is guided into the integrated planar waveguides by optical fibers glued onto the endface of the chip. Such a device can in principle be connected to the light source and photodetector in a 'plug-and-play' manner.

In a similar approach, optical fibers were inserted into etched grooves and used to illuminate and collect light from a small channel segment. Here, integration of planar waveguides is avoided, but optical fibers allow detection at only a single point per fiber, in contrast to planar waveguides, which can be branched into many detection points.

In both approaches the light beam passes unguided through the detection channel, and a major fraction of the light is typically lost, because it is not collected by the waveguide or optical fiber on the other side of the channel. It propagates as stray light in the plane of the device or radiates out of the chip. This problem has been addressed by fabrication of so-called evanescent-wave sensors, in which the beam is also guided in the detection region.

4.4.3
**Evanescent-wave Sensing**

In a planar waveguide or in an optical fiber, all the optical power is not confined in the core layer, but a fraction of the electromagnetic field decays exponentially to zero outside the waveguide core region. This area is called the evanescent field region (Fig. 4.28) and can be several micrometers thick, depending on the refractive index of the various layers and the wavelength of the light.

Normally, a so-called waveguide-cladding layer with a low refractive index is deposited on top of the high-index waveguide core layer to prevent interference from the external environment. But for chemical microsystems it is possible to take advantage of this interference by using it for sensing. This principle has been used in a device consisting of a branching planar waveguide integrated with a microliquid handling system [48] (Fig. 4.29):

A waveguide branches into a sensing arm and a reference arm. In the sensing arm, a region of the waveguide-cladding layer is removed, so the evanescent field is exposed to the sample. Evanescent-wave sensing can be performed in several ways. The simplest approach is based on absorption measurement in which a wavelength is chosen that corresponds to an absorption peak of the sample. This results

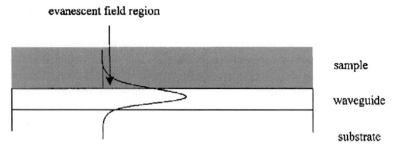

**Fig. 4.28** Principle of evanescent-wave sensing. The evanescent field that extends outside the waveguide core is used for probing the sample [48].

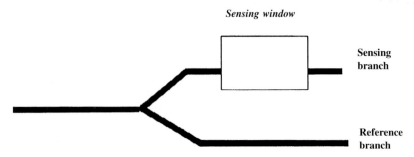

**Fig. 4.29** Configuration of the optical system in a device based on evanescent-wave sensing. The sensing window is 1000 μm long.

in attenuation of the optical power in the sensing arm compared to the reference arm, which can be described by the Lambert–Beer law (Eq. 4.8 and Eq. 4.9).

Evanescent-wave sensing of nonabsorbing samples can also be done by measuring the change in refractive index, which also depends on the solute concentration. This approach relies on measurements of the interference pattern between the sensing branch and the reference branch. It can be orders of magnitudes more sensitive than traditional absorbance measurements, but is unfortunately also difficult to control, because other factors such as the temperature influence the refractive index of the sample. The optical readout is also more complicated.

### 4.4.4
### Fluorescence Detection

Fluorescence spectroscopy involves excitation of an electron to a higher-energy state due to absorption of a photon (see section 4.4.2), followed by relaxation of the electron to the lower-energy states (Fig. 4.30) resulting in emission of photons in addition to dissipation of thermal energy to the surroundings. When a laser is used for excitation, the detection method is called laser-induced fluorescence (LIF).

The emission spectrum is always shifted towards longer wavelengths than the absorption spectrum (Fig. 4.31), because energy is conserved throughout the excitation and relaxation processes. Because a fraction of the excitation energy is typically dissipated as thermal energy (nonradiative relaxation), less energy is left for emission of photons (radiative relaxation); thus, the emission spectrum is located at lower energies than the absorption spectrum. This corresponds to a longer wavelength, because energy and wavelength are inversely proportional, as seen in the well-known formula:

$$E = hv = \frac{hc}{\lambda} \tag{4.10}$$

For calculations it is easier to remember the formula as

$$E[eV] = \frac{1.24}{\lambda[\mu m]} \tag{4.11}$$

where energy is in units of electron volt and wavelength is in micrometers.

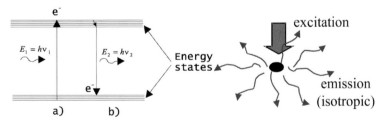

**Fig. 4.30** Left: Energy diagram showing the concept of fluorescence. A photon with energy $E_1$ is absorbed, and a photon with lower energy $E_2$ is subsequently emitted. Right: Schematic drawing showing the isotropic nature of the emitted light.

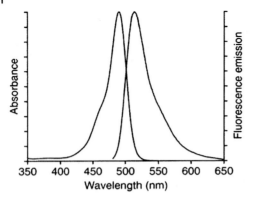

**Fig. 4.31** Absorption (left) and emission (right) spectra of fluorescein, a widely used fluorophore.

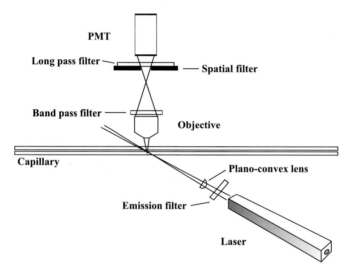

**Fig. 4.32** Schematic drawing of a typical optical setup for fluorescence measurements [49].

Fluorescence measurement is widely used because very low limits of detection can be achieved – it is even possible to measure single molecules. A disadvantage compared to other techniques such as absorption measurement is that the optical setup is relatively complex, because the emitted light has to be separated from the excitation light, since collection of excitation light results in an increase in the background signal and hence an increase in the noise of the measurement. A schematic drawing of a typical setup is shown in Fig. 4.32.

A laser is focused onto a section of a fluidic channel from an angle of about 45° to avoid shining its light directly into the photodetector. The fluorescence is collected by an objective lens before which filters, such as bandpass and long wave pass filters, are inserted in the light path to suppress collection of excitation light. A spatial filter is also typically used to ensure that only light from a well-de-

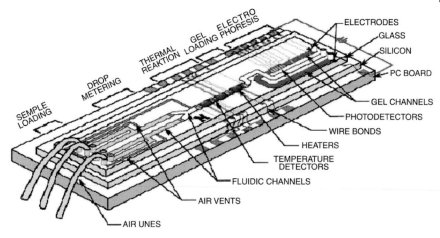

**Fig. 4.33** Microfabricated device for DNA analysis with an integrated fluorescence detector [51].

fined region of the channel is collected. Hence, many different optical components are used. The setup is also bulky and very sensitive to correct alignment of the excitation light and the collection optics.

One of the most impressive devices in terms of integration was developed for fluorescence measurements in DNA analysis [50] (Fig. 4.33). This device was fabricated from a silicon substrate with integrated photodiodes for optical detection. A thin film interference filter was deposited on top of the diode to suppress measurement of excitation light. Optical excitation was applied from the top of the device with an external blue-light-emitting diode (LED). The fluidic channel network consisted of a sample loading compartment, a PCR chamber for DNA amplification, and channels for separation of the DNA molecules by capillary electrophoresis.

## 4.5
## Electrochemical Sensors

Electrochemical sensors can often be applied if the sample is for instance turbid. Another advantage is that electrochemical sensors are relatively easy to fabricate because they require only two electrodes that normally consist of noble metals such as platinum, silver, or gold. Other electrode materials used are vitreous/ glassy carbon or graphite or, more advanced because optically transparent, indium tin oxide (ITO). The measurement principles most often applied are potentiometry and amperometry. As the name indicates, a potential or a current is measured while keeping the current flow or the potential difference between two electrodes constant.

One important aspect of each electrochemical measurement is to chose and fabricate the right reference electrode. The purpose of a reference electrode is to have a fixed potential to relate the signal to, because only potential differences can be measured. Conventionally, all redox potentials were related to the normal hydrogen electrode (NHE), an electrode that consists of a platinum mesh, dipped into 1 M sulfuric acid, purged with 1.013 bar hydrogen at a temperature of 25 °C. The potential of the NHE was set to 0 V.

Because this type of electrode is not very convenient, other stable reference electrodes were invented. Common, so-called second-generation, reference electrodes consist of a metal such as mercury or silver and a corresponding, barely-soluble salt such as $Hg_2Cl_2$ (calomel) or AgCl. The purpose is to create a saturated solution of the ions, resulting in a stable reference potential. However, this reference potential is still temperature dependent, for instance, the potential of a Ag/AgCl system with saturated KCl solution is 220.5 mV at 0 °C and 159.8 mV at 60 °C [51].

Because calomel electrodes are somewhat environmentally problematic, Ag/AgCl is used much more today. It is actually unimportant which system is used – both types have the same quality.

Several researchers have tried recently to fabricate stable and accurate miniaturized reference electrodes; however, most were unstable or inaccurate. The major flaw lies in the 'consumption' of either the barely-soluble salt or the electrode material. A thin-film electrode consisting of, for instance, only 1000 nm Ag with a surface area of less than a square millimeter completely dissolves within a couple of weeks according to the reaction:

$$Ag \rightarrow Ag^+ + e^- \tag{4.12}$$

and a small volume of solid AgCl dissolves very quickly if washed with deionized water for even a short period. This of course affects the stability of that reference electrode, resulting in a drifting and most often unstable signal.

However, microscale Ag/AgCl systems intended for short-term use (<2 weeks working condition) can be built by relatively simple techniques: an Ag electrode (e.g., thin film Ag deposited on a silicon chip) is dipped into a chloride solution and covered electrochemically at potentials of $\sim 150$ mV against a commercial Ag/AgCl reference electrode, or AgCl can be deposited out of an iron(III) chloride solution according to the following equation:

$$FeCl_3 + Ag \rightarrow AgCl + FeCl_2 \tag{4.13}$$

Slight modifications of the electrode setup offer a variety of analytical possibilities – the so-called ion-selective electrodes (ISE). As an example, the fabrication of a potassium-selective electrode is given in section 10.1.

To miniaturize potentiometric sensors, several attempts have been made. One of the most elegant techniques for making miniaturized electrochemical systems is based on the ion-selective field effect transistor (ISFET). Bergfeld et al. [52] removed the metal oxide of a metal oxide field effect transistor (MOSFET) and replaced it with an ion-selective membrane mix. In general, the functional principle of an ISFET is similar to that of a triode. Triodes were used around the 1960s in radios and later in televisions.

In a *p*-type doped silicon wafer, two regions on the top are doped with electron donators such as nitrogen or phosphorus. Between those two regions, called source (S) and drain (D) of the FET a thin insulating layer is deposited. Commonly used materials are silicon oxide or silicon nitride. If the FET is connected as in Fig. 4.34, a thin electron channel between source and drain is developed right at the interface between the insulator and the bulk *p*-doped silicon.

Depending on the potential difference between the gate and the bulk silicon, this electron channel becomes more or less deep, thereby influencing its conductance. This is known as the field effect. Because the conductance change itself influences current flow between source and drain, this current flow is directly proportional to the gate potential. In principal, an FET with a gate made of silicon nitride or tantalum oxide is sensitive towards protons and can therefore be considered an ISFET that senses the pH value. If the gate is now covered with an ion-selective membrane, other ions such as ammonium, potassium, sodium, nitrate, and many others can be measured.

The first commercial ISFETS were pH-sensitive due to a thin layer of tantalum oxide or silicon nitride – both materials are frequently used as masking materials in silicon processing and therefore are well known to processing engineers. An additional advantage is the adhesion of these inorganic layers: they stick to the gate surface much better than the above-mentioned liquid membranes.

After the first pH-ISFETS entered the market, several ISFET-type sensors were developed and commercialized. The 'classic' pH-ISFET is sold as an 'unbreakable' alternative to the glass electrode. However, because it is machined of crystalline silicon, this material also can break, and additionally, measurement data obtained from ISFETs tend to drift. The main application of pH-ISFETs is in the food industry: the sensor can for example be inserted directly into an apple or a tomato – an advantage compared to the thin glass membrane of a glass electrode, which would break immediately. The advantage of being small with respect to the glass electrode has, surprisingly enough, not resulted in many commercial products yet. The reason for this is most likely one of the still unsolved problems in elec-

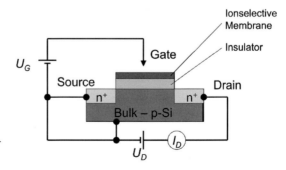

**Fig. 4.34** Schematic drawing of ion-selective field effect transistor (ISFET) [53].

trochemistry: constructing a functional miniaturized reference electrode. One approach to circumvent the necessity of a reference electrode so as to take advantage of miniaturized sensors is to use differential techniques for the measurement. Instead of a reference electrode, a second so-called reference-FET or ReFET can be used. However, it has to be demonstrated that both FETs (ISFET and ReFET) have the same slope and drift in signal.

## 4.6
## References

1 Kovacs, G.T.A. *Micromachined transducers sourcebook*; McGraw-Hill: New York, 1998.

2 Woias, P. *Proceedings of SPIE*, San Francisco 2001; SPIE; 39–52.

3 Shoji, S. In *Microsystem Technology in Chemistry and Life Science*; Becker, H., Manz, A., Eds., 1998; Vol. 194, pp 164–188.

4 Madou, M. *Fundamentals of microfabrication*, first ed.; CRC Press: Boca Raton, Florida, 1997.

5 Yu, Q.; Bauer, J.M.; Moore, J.S.; Beebe, D. J. *Applied Physics Letters* **2001**, *78*, 2589–2591.

6 Beebe, D. J.; Moore, J.S.; Bauer, J.M.; Yu, Q.; Liu, R.H.; Devadoss, C.; Jo, B.H. *Nature* **2000**, *404*, 588–590.

7 Krog, J.P.; Dirac, H.; Fabius, B. et al. Proceedings of the μTAS 2000 Symposium, Enschede, Netherlands, May 14–18, 2000; Kluwer Academic Publishers; 419–422.

8 Duffy, D.C.; Gillis, H.L.; Lin, J.; Sheppard, N.F., Jr.; Kellogg, G.J. *Analytical Chemistry* **1999**, *71*, 4669–4678.

9 Ekstrand, G.; Holmquist, C.; Örlefors, A.E.; Hellman, B.; Larsson, A.; Andersson, P. *Micro Total Analysis Systems 2000: Proceedings of the μTAS 2000 Symposium*, Enschede, The Netherlands, May 14–18, 2000; Kluwer Academic Publishers; Dordrecht; 311–314.

10 Thomas, N.; Ocklind, A.; Blikstad, I.; Griffiths, S.; Kenrick, M.; Derand, H.; Ekstrand, G.; Ellström, C.; Larsson, A.; Andersson, P. *Proceedings of the μTAS 2000 Symposium*, Enschede, The Netherlands, May 14–18 2000; Kluwer Academic Publishers; Dordrecht; 249–252.

11 Adamson, A.W. *Physical Chemistry of Surfaces*; John Wiley and Sons: New York, 1990.

12 Schwer, C.; Kenndler, E. *Analytical Chemistry* **1991**, *63*, 1801–1807.

13 Li, S.F.Y. *Capillary Electrophoresis*; Elsevier: Amsterdam, 1993.

14 Paul, P.H.; Garguilo, M.G.; Rakestraw, D.J. *Analytical Chemistry* **1998**, *70*, 2459–2467.

15 Guo, Y.; Imahori, G.A.; Colon, L.A. *Journal of Chromatography A* **1996**, *744*, 17–29.

16 Barker, S.L.R.; Tarlov, M. J.; Canavan, H.; Hickman, J.J.; Locascio, L.E. *Analytical Chemistry* **2000**, *72*, 4899–4903.

17 Barker, S.L.R.; Ross, D.; Tarlov, M.J.; Gaitan, M.; Locascio, L.E. *Analytical Chemistry* **2000**, *72*, 5925–5929.

18 Schasfoort, R.B.M.; Schlautmann, S.; Hendrikse, J.; van den Berg, A. *Science* **1999**, *286*, 942–945.

19 Ramsey, R.S.; Ramsey, J.M. *Analytical Chemistry* **1997**, *69*, 1174–1178.

20 Lazar, I.M.; Ramsey, R.S.; Jacobson, S.C.; Foote, R.S.; Ramsey, J.M. *Journal of Chromatography A* **2000**, *892*, 195–201.

21 Culbertson, C.T.; Ramsey, R.S.; Ramsey, J.M. *Analytical Chemistry* **2000**, *72*, 2285–2291.

22 McKnight, T.E.; Culbertson, C.T.; Jacobson, S.C.; Ramsey, J.M. *Analytical Chemistry* **2001**, *73*, 4045–4049.

23 Alarie, J.P.; Jacobson, S.C.; Broyles, B.S.; McKnight, T.E.; Culbertson, C.T.; Ramsey, J.M., Monterey, California, USA 2001; Kluwer Academic Publishers; 131–132.

24 Paul, P.H.; Arnold, D.W.; Rakestraw, D.J., Banff, Canada 1998; Kluwer Academic Publishers; 49–52.

25  PAUL, P.H.; ARNOLD, D.W.; NEYER, D.W.; SMITH, K.B. In *µTAS 2000*; VAN DEN BERG, A., OLTHUIS, W., BERGVELD, P., Eds.; Kluwer Academic Publishers: Monterey, California, 2000, pp 583–590.

26  ZENG, S.; CHEN, C.-H.; MIKKELSEN, J.C.; SANTIAGO, J.G. *Sensors and Actuators B* **2001**, *79*, 107–114.

27  STROOCK, A.D.; DERTINGER, S.K.W.; AJDARI, A.; MEZIC, I.; STONE, H.A.; WHITESIDES, G.M. *Science* **2002**, *295*, 647–651.

28  JOHNSON, T.J.; ROSS, D.; LOCASCIO, L.E. *Analytical Chemistry* **2002**, *74*, 45–51.

29  ODDY, M.H.; SANTIAGO, J.G.; MIKKELSEN, J.C. *Analytical Chemistry* **2001**, *73*, 5822–5832.

30  JACOBSON, S.C.; KOUTNY, L.B.; HERGENRÖDER, R.; MOORE JR., A.W.; RAMSEY, J.M. *Analytical Chemistry* **1994**, *66*, 3472–3476.

31  JACOBSON, S.C.; HERGENRÖDER, R.; MOORE JR., A.W.; RAMSEY, J.M. *Analytical Chemistry* **1994**, *66*, 4127–4132.

32  ALARIE, J.P.; JACOBSON, S.C.; CULBERTSON, C.T.; RAMSEY, J.M. *Electrophoresis* **2000**, *21*, 100–106.

33  ZHANG, C.X.; MANZ, A. *Analytical Chemistry* **2001**, *73*, 2656–2662.

34  LAURELL, T.; WALLMAN, L.; NILSSON, J. *Journal of Micromechanics & Microengineering* **1999**, *9*, 369–376.

35  PETERSSON, M.; NILSSON, J.; WALLMAN, L.; LAURELL, T.; JOHANSSON, J.; NILSSON, S. *Journal of Chromatography B.* **1998**, *714*, 39–46.

36  U.D. LARSEN, *Micro Liquid Handling*, PhD Thesis, Mikroelektronik Centret, Kgs. Lyngby, Denmark, July 2000.

37  C.J. YEAGER and S.S. COURTS, *IEEE Sensors Journal*, **2001**, 1(4), 352–560.

38  A.F.P. VAN PUTTEN, *Electronic Measurement Systems*, 2nd edit, Institute of Physics Publishing, Bristol, **1996**.

39  Z. ZHAN, C. DAFU, Y. ZHONGYAO, W. LI, *Biochip for PCR amplification in silicon*, **2000**, 1st Annual International IEEE-EMBS Special Topic Conference on Microtechnologies in Medicine & Biology, Lyon, October 12–14.

40  P. HAUPTMANN, *Sensoren: Prinzipien und Anwendungen*, Carl Hanser, Munich, **1990**.

41  R. FRANK, *Understanding Smart Sensors*, Artech House, Boston, **1996**.

42  Y. SATO, G. IRISAWA, M. ISHIZUKA, K. HISHIDA and M. MAEDA, *Measurement Science and Technology*, **2003**, 14, 114–121.

43  S.W. ALLISON and G.T. GILLIES, *Review of Scientific Instruments*, **1997**, 68 (7), 2615–2650.

44  A.M. CHAUDHARI, T.M. WOUDENBERG, M. ALBIN and K.E. GOODSON, *Journal of Microelectromechanical Systems*, **1998**, 7(4), 345–355.

45  G.C.M. MEIJER and A.W. VAN HERWAARDEN, ED, *Thermal Sensors*, Institute of Physics Publishing, Bristol, **1994**.

46  R. MARIE, J. THAYSEN, C.B.V. CHRISTENSEN and A. BOISEN, *Microelectronic Engineering*, *1*, **2003**, 67–68, 893–898.

47  K.B. MOGENSEN et al. *Electrophoresis*, 22 (2001), p 3930.

48  G. PANDRAUD et al. *Sens. Actuat. A*, 85 (2000), p 158.

49  S.A. SHIPPY et al., *Analytica Chemica Acta* 307 (1995).

50  BURNS et al., *Science*, **1998**.

51  H.R. CHRISTEN, *Grundlagen der allgemeinen und anorganischen Chemie*, Verlag Sauerlaender Ag Aarau, 9. Auflage **1988**.

52  P. BERGVELD, *Ion sensitive field effect transistor*, International Conference on Biomedical Transducers, **1975** pp 299–304.

53  J.K. CAMMANN, *Instrumentelle Analytische Chemie*, **2000**, Spektrum Akademischer Verlag, Heidelberg, Berlin.

# 5
# Simulations in Microfluidics

Goran Goranovic and Henrik Bruus

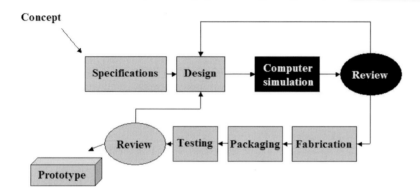

This chapter shows by example how we can use computer simulations in designing and testing microfluidic systems. The fabrication of µTAS systems is very complex and requires interdisciplinary research and development. Moreover, the production time in a cleanroom facility of, say, a silicon-based µTAS can be several weeks or even months. Numerical simulations of any given chip design are therefore extremely useful. Not only can they provide a more complete understanding of the fundamental physical and chemical processes of the entire µTAS, but they can also be used to develop optimal designs and to minimize the risk of wasting expensive production time on a flawed design.

It is by no means possible to cover or even mention all relevant aspects of simulations in microfluidics in a short overview. Therefore, descriptive explanations prevail over details in an attempt to help you visualize a helpful, 'zeroth-order approximation'.

The structure of this chapter itself reflects how we can deal with simulations by working through various phases: analyzing the fundamental physics and chemistry of a given µTAS, choosing the right software and hardware, setting up a simulation, analyzing the sources of numerical errors and uncertainties, and finally, evaluating the precision and accuracy of the results. Indeed, the importance of the last point cannot be overrated. It is nearly always possible to generate results, but are they correct and do they describe the actual device?

*Microsystem Engineering of Lab-on-a-chip Devices*
O. Geschke, H. Klank, P. Telleman
Copyright © 2004 Wiley-VCH Verlag GmbH & Co. KGaA, Weinheim
ISBN: 3-527-30733-8

## 5.1
## Physical Aspects and Design

When designing new microfluidic devices, the designer must have a proper understanding of the physical aspects of the problem. Some of these are listed next.

**Dimensions**  A microfluidic system is a network of fluidic channels and other components such as valves and pumps. It typically occupies an area of one square centimeter, with channel widths typically on the order of 100 μm. For comparison, a human hair has a thickness of approximately 50 μm, and the diameter of red blood cells is about 7 μm. From a practical point of view, even smaller dimensions in a lab-on-a-chip are not necessarily advantageous, because the requirement for easier handling as well as measurements can impose some restrictions. However, if the goal is to develop systems for single-molecule detection, the submicrometer is unavoidable, and one enters the new and exciting field of nanofluidics.

**Geometry**  The basic component of a microfluidic network is a channel. Its main features include length, cross-section, and surface properties such as roughness. After a channel is made in a substrate, it is covered with a bonded lid that can be made of a different material.

Long channels are usually needed if a reaction of several mixed chemicals is to occur (in laminar-flow regime, diffusion is the main mixing mechanism, which takes time). Also, when a chemical compound consisting of several components needs to be analyzed, usually by electrical separation techniques, the separation

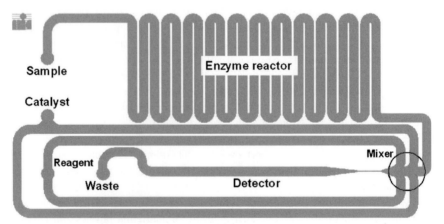

**Fig. 5.1**  Mask of an integrated microsystem for detecting chemical reactions by chemiluminescence. Before entering the mixing region, a sample passes through a long spiral for a prolonged reaction with immobilized enzyme. Then, a product, usually $H_2O_2$, enters the mixing region and, via a catalyzed reaction with reagent, produces light. Photodiodes, used for the light detection, are on the back and not shown. The chip dimensions are 20 mm×10 mm, and the width of the enzyme reactor channel is 400 μm. Courtesy A. M. Jorgensen.

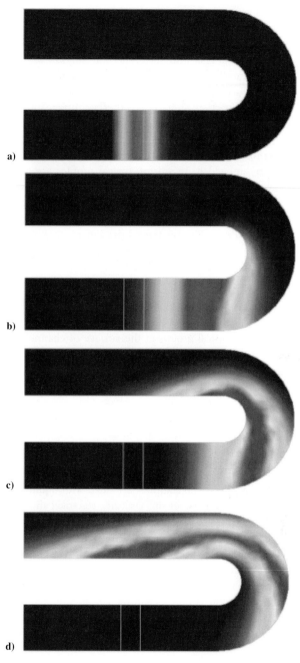

**Fig. 5.2** Race-track effect in a turn in an electrokinetically driven flow. The concentration contours of a sample are displayed. The inside track has a shorter path, as well as a higher electrical field, inducing faster movement of the inner molecules and therefore enhancing sample dispersion.

channel is made longer to allow for proper separation of the components. Because microchips are confined to a small area, the extra length is achieved by meander or spiral structures, Fig. 5.1). Such channel bending, however, induces dispersion of a chemical species as the molecules travel different distances inside the turn (race-track effect, Fig. 5.2). Dispersion decreases the concentration, resulting in reduced resolution, i.e., larger overlaps of the concentration peaks and diminished detection signals. Another common source of geometrical dispersion are interconnections, such as tubing, between a chip and external liquid reservoirs.

Channels can have different cross sections, depending on the material in which they are embedded. In silicon, typical profiles are rectangular, or actually slightly trapezoid due to underetching. In polymers, laser beams can produce rectangular, triangular, and gaussian-like shapes. Circular cross sections, as in capillaries, are also encountered. In addition, channels can have various degrees of surface roughness depending on the fabrication process. Geometrical features are important because they determine the characteristic resistances of a channel. The hydraulic resistance, a concept used in the uniform flow regime, relates an applied pressure to the corresponding flow rate, and the electrical resistance of a channel filled with a conducting liquid relates an applied voltage to the passing current.

**Surface**   Chemical groups at the surface of channel walls, such as silanol (SiOH) groups in glass, can react with the ions in an electrolyte solution and create a very thin polarized layer. This electric double layer, or Debye layer, has a thickness of 1–10 nm and is responsible for electroosmotic flow (EOF) in an applied electric field. Surface double layers are characterized by the zeta potential. Different surfaces, such as the lid and the bottom of a channel, can have different zeta potentials. The characteristic EOF velocity depends on the zeta potential through the electroosmotic mobility, which is largely an empirical value.

**Fluid properties**   The channels are filled with liquids that can have different values of properties such as density, viscosity, electrical and thermal conductivity, diffusion coefficients, and surface tension. Often a system has two types of liquids – a buffer and a sample. The buffer is usually the transporting liquid resisting the changes in pH value that can be, for example, induced close to electrodes or in chemical reactions. A sample is a liquid of interest that needs to be analyzed. Both buffer and sample are usually placed on the chip via separate reservoirs, with the buffer finally surrounding the sample. A sample confined to a specific region is called a sample plug. In separation science, the sample consists of several chemical species, such as a positively charged, a negatively charged, and a neutral species. As mentioned above, the initial concentration of the sample plug is decreased by dispersion in the system. Besides geometrical constrictions, velocity profiles, such as the parabolic in pressure-driven flows, and diffusion also cause dispersion.

Fluids such as blood or gels are so-called non-Newtonian liquids, for which the viscosity is not constant but is a function of the local shear stress. In [3] a simple power-law model for blood viscosity is treated, and a variety of other models are summarized. The resulting flow profile has a blunted rather than the usual parabolic form.

***Heating*** Joule heating is generated when an electrical current flows through a channel. In some applications, such heating may destroy biological samples. It may also induce changes in viscosity and thus affect the velocity profile. Moreover, due to enhanced Brownian motion, heating may increase the dispersion of a sample plug. Heating may also increase the mobility of the ions in electric double layers, leading to enhanced flow rates in electroosmotic flows [5]. Finally, heating stimulates production of bubbles of the gases dissolved in the liquid. The bubbles are usually unwanted and can block flow if they get stuck in narrow contractions.

For a given application only some phenomena need to be taken into account. In many microfluidic simulations, heating is often neglected in the first considerations.

## 5.2
## Choosing Software and Hardware

When choosing a commercial computational fluid dynamics (CFD) package, there are basically two guidelines: its capability (including the company's support) and its price.

Most companies give a one-month free trial so that a user can get a feeling for the tool. However, because the tools are rather comprehensive, one month is often not enough time. Therefore, it is a good practice to think of an instructive, fairly simple test case (different from the tutorials) that can be easily checked in the literature. If possible, several software tools ought to be evaluated on the same problem so they can be compared directly.

Software companies generally distinguish between academic licenses (a few thousand euros), aimed at universities, and commercial licenses (several tens of thousands of euros), aimed at companies. Some companies give special inexpensive annual academic licenses (a couple of thousands of euros) that are extremely attractive. These can be renewed every year for an even lower cost.

Of all CFD packages available today on the market, two are discussed more thoroughly because they were used to produce the simulations in this chapter.

### 5.2.1
### CFD-ACE+Version 6.6

This software is a comprehensive finite-volume tool with an intuitive interface [1]. The original version was designed for macroscopic CFD problems (including turbulence), but later, extensive modules have been added to describe a more general range of problems including those of microtechnology. At present, it includes about 20 different modules. In microfluidics, specific combinations of flow, heat transfer, chemistry, free surface, electric, magnetic, and biochemistry modules are used to tackle phenomena such as pressure-driven flows, transport of numerous chemical species (e.g., electrophoresis), electroosmosis, bubble behavior, binding of biomolecules, etc.

The program has excellent grid generation capabilities, supporting both structured and unstructured grids. It includes scripts written in the Python language, which makes optimization of geometry as well as parametric runs (i.e., running consecutive simulations, changing one or more parameters) very easy to handle.

The numerical tools include a variety of boundary conditions, various discretization schemes, etc.; even parallel processing is possible. Numerous ways of viewing the results are available. One of the strengths of the tool is its extensive manuals in which everything from numerics to the physical governing equations is described in detail. In particular, limitations of the models are discussed. The company CFDRC, Huntsville, AL, USA, delivers good professional support, is open to collaboration, and offers annual academic licenses.

Simulations often require long run times. Fortunately, a new solver can be selected to speed up convergence. A solver is a particular algorithm used to obtain numerical solution (see Sections 5.2.6 and 5.3.2). Extra attention needs to be paid to the properly set of numerical parameters, which requires a knowledgeable user. This, however, is beneficial in the long run, because the user has control over the calculations. The program can run on Windows, Unix, and Linux. It requires about 320 MB of hard drive space for installation of its standard package. More on specifics of the package can be found in [2,3], where bubble motion in microchannels and non-Newtonian microflows are considered.

## 5.2.2
## CoventorWare™ Version 2001.3

This software package is specifically oriented to microtechnology (to both microelectromechanical systems (MEMS) and microfluidics) [4]. CoventorWare™ has no turbulence module, which is probably unnecessary in microfluidics anyway. Another characteristic feature is that the grid generation emulates the manufacturing process, which automatically guides the user in making realistic designs. The possible designs are two-dimensional, so more complicated structures and grids need to be designed with an external program such as I-DEAS® and subsequently imported.

The CoventorWare™ package contains many different modules. Three are of particular importance in microfluidics (parts of Analyzer™): MemCFD covers general CFD tasks (steady, transient, compressible, and incompressible laminar flows; suitable for pumps and valves); NetFlow covers pressure, diffusion, electrophoresis, electroosmosis, and chemical transport of up to four species; and SwitchSim has very convenient cyclic voltage boundary conditions useful when studying time-dependent separations on a chip. The results can be visualized in a variety of ways.

A strong point of the program is that the solvers are fast (NetFlow uses another CFD program called Fluent®, which is well tested). Also, different parametric runs can be implemented easily. Another attractive feature is that the system-level modeling, based on the equivalent circuit theory, for more complicated fluidic networks will soon be incorporated in a user-friendly manner. The downsides are that in many aspects CoventorWare™ is like a black box, without sufficient explanation in the manuals of the working mechanisms, both numerical and theoretical. Also, complex interfacing of the external solvers can lead to some compatibility errors. However, the support staff of Conventor Inc., Cary, NC, USA, is exceptionally professional, and all requests and questions are answered in a detailed, timely manner. An additional point is that the user interface is somewhat non-standard compared to other CFD programs. CoventorWare™ runs on Windows and Unix. It uses about 380 MB of hard-drive space for installation and requires at least 512 MB of RAM to run properly, which is rather demanding. For a thorough overview of the program with excellent examples of electroosmotic pumps and dispersion phenomena, please see [5, 6].

To conclude: CFD-ACE+ and CoventorWare™ are generally good and useful, with specific advantages and limitations. In the end, it is the user's knowledge that makes a difference. Apart from being prone to make more frequent mistakes, untrained users cannot exploit such extensive software to the full extent.

Standard mathematics programs, such as MATLAB® [7] and MATHEMATICA® [8], are useful supplements in computing analytical (exact) solutions of equations for comparison with approximate CFD simulations (the latter often do not model all aspects of interest).

### 5.2.3
### Hardware

The simulations presented later in this chapter were done on PCs running the Windows 2000 operating system and having these specifications:

| | |
|---|---|
| processor: | Intel Pentium III processor, 933 MHz |
| memory: | 512 MB |
| video: | Intel 82815 graphics controller (AGP) 4 MB of memory |
| hard drive: | 40 GB |
| other: | 17″ monitor, 3-button mouse, a CD writer (for large simulations) |

### 5.2.4
### The Core Elements of Typical CFD Software

Any CFD package typically consists of a pre-processor, a solver, and a post-processor. These elements are briefly described below.

### 5.2.5
### Pre-processors

A pre-processor is the interface between the user and the numerical solver. Several factors must be considered before the solver can be started.

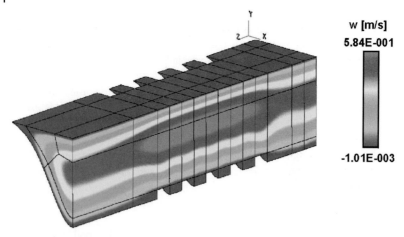

**Fig. 5.3** Computational domain of a 3D channel laser-machined in plastic. Because of symmetry, only half the full domain needs to be simulated. The geometry includes a characteristic gaussian-like cross-section and perpendicular ripples, which can arise from the fabrication process. The contours show the axial velocity profile [9]. Performed on CFD-ACE 6.6.

***Geometry*** In solving CFD problems the first step is to define a computational domain, i.e., the region in which the fluid equations are being solved. Using symmetry arguments can often reduce the region substantially, thus decreasing computational time. Therefore, the computational domain should be the minimum region needed to be simulated (Fig. 5.3). For three-dimensional (3D) structures it is a good idea to first simulate similar and less time demanding two-dimensional (2D) problems until all the parameters are properly set. Some programs offer parametric specification of geometry, which allows the complicated designs to be easily altered.

***Grid generation*** To obtain appropriate results the computational domain needs to be divided into a number of cells by using a grid or mesh. In general, finer grids yield more accurate results, but also increase the computational time immensely. Thus it is very important to generate an efficient grid.

The two most common types of grids are structured and unstructured. Structured grids are more regular in the sense that neighboring cells can be easily counted (Fig. 5.4). They usually include four-sided polygons such as rectangles (2D) or bricks and prisms (3D). A well structured grid includes high orthogonality, low skewness, and low-to-moderate aspect ratios of the cells. An unstructured grid can more easily be adapted to irregular geometries because it includes triangles in 2D and tetrahedra in 3D. Its generation and implementation, however, require more complicated algorithms and can lead to increased numerical diffusion, e.g., in shear regions [10]. Numerical diffusion is the accumulation of numerical rounding-off errors due to discretization of the differential equations (Section 5.3). A hybrid grid combining both structured and unstructured grids is shown in Fig. 5.5.

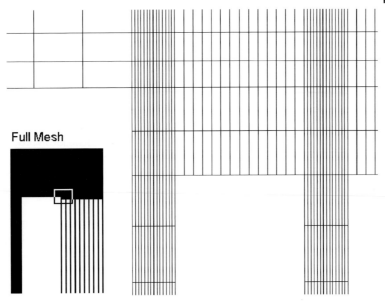

Full Mesh

**Fig. 5.4** Example of a structured grid (top view) used in simulations of electroosmotic pumps [5]. Of the three cell sizes, the finest are used to resolve velocity profiles in narrow channels. In the upper left, large cells are adjacent to small cells, and
computations cannot be trusted near the junctions. Experience has shown that integrated quantities such as flow rate and pressure are less affected by grid quality
than velocity profiles. Calculated with CoventorWare™ 2001.3. Courtesy of A. Brask.

**Fig. 5.5** A hybrid grid combining
structured and unstructured parts.
The domain presents a quarter of a
2D plate with a hole in the middle.
The region near the hole is meshed
regularly, and farther away the grid
becomes unstructured. The grid is
denser in the regions of expected
stress. Calculated with CFD-ACE+ 6.6.

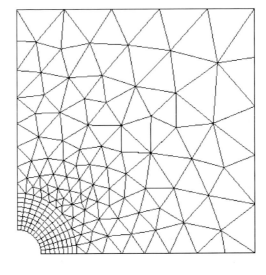

The solution should be independent of the grid, which can be checked by monitoring the solution as a function of grid refinement. Often the grid is refined successively by a factor of two until the change in calculated quantities is smaller than a user-defined tolerance. Regions in which large gradients in velocity or concentration are expected require a more highly resolved grid. In particular, areas around sharp corners need to be well resolved. The grid aspects are discussed further in the test cases.

**Model** The model of the physical, electrical, and biochemical processes to be simulated needs to be defined with care. Sometimes a simple model with clear physical implications may lead to better understanding of the resulting CFD simulations than a model of unnecessary complexity.

**Fluid parameters** Parameters such as viscosity, diffusion coefficients (Tab. 5.1), and electroosmotic mobility need to be specified. The relevant data can be extracted from the software databases or can be found in literature, one good source being [11]. However, sources in the literature do not cover the ever-increasing number of materials used in new applications. For example, the diffusion coefficient of a DNA strand depends on its length and the liquid environment and obviously cannot be arbitrarily given. New sets of experiments might be needed to provide the correct input data, and you may need to search for scientific publications dealing with the subject.

**Boundary conditions (BCs)** The BCs are specified on the surfaces that define the computational domain. Typical boundary conditions are velocity, pressure, wall effects (EOF), voltage, and so forth. A more detailed overview of BCs is given in the following section.

**Initial conditions (ICs)** The ICs are the values of the variables at an initial time. When searching for a steady-state solution, a good guess of ICs reasonably close to the solution is often needed to shorten computational time and to avoid divergence. When solving time-dependent problems, the late-time solutions of one simulation may serve as ICs for subsequent simulations.

**Tab. 5.1** Some mass diffusion coefficients at 25 °C, [11].

| Species | $D_{mass}$ (m²/s) |
|---|---|
| Water | $0.890 \times 10^{-6}$ (self-diffusion) |
| Ethanol | $1.24 \times 10^{-9}$ (in water) |
| Glucose | $0.67 \times 10^{-9}$ (in water) |
| Sucrose | $0.52 \times 10^{-9}$ (in water) |
| Urea | $1.38 \times 10^{-9}$ (in water) |

***Solver settings*** Numerical parameters are adjusted so that the solvers can solve the problem at hand most efficiently. These settings include type of solver, numerical schemes, number of iterations, convergence criteria, relaxation parameters, etc. A more detailed description, along with some definitions, is given in the following section as well.

### 5.2.6
### Solvers

Solvers are numerical algorithms intended to solve the governing equations. The following two methods are most common in commercial CFD programs.

***Finite-volume method*** The finite-volume method (FVM) is the most widely used CFD technique. It is based on the concept of transport equations (Fig. 5.6). The fluid domain is divided into finite volumes, and the governing equations are integrated over these volumes. The integrated equation is approximated by various finite-difference methods such as the central difference and the upwind scheme. The numerical schemes convert the integral equations into a system of algebraic equations. These equations are then solved by an iterative method. Specific solver properties are discussed in Section 5.3.2.

***Finite-element method*** The finite-element method (FEM) makes use of simple piecewise functions, such as linear or quadratic functions defined on small elements, to approximate the exact solution. A residual is defined to measure how far the approximate solution is from fulfilling the governing equations. In the so-called Galerkin method, in an iterative procedure these residuals are then minimized by using them to correct the approximate solution [12].

FVM is used mostly for fluid dynamics, and FEM is more suitable for mechanical stress analysis and simulations of MEMS. The FEM method usually uses an unstructured grid, and the grid-quality requirement is less stringent than for FVM. On the other hand, FEM requires more computer memory than FVM on structured grids with same number of nodes [13].

### 5.2.7
### Post-processors

A post-processor is used to visualize results. Some commonly used visualization techniques follow.

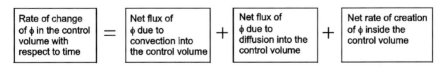

**Fig. 5.6** Description of the transport equation used in the finite volume method.

***Translation/rotation/scaling*** Computational domains can be easily manipulated, probed, and viewed from many sides that would otherwise, as in experiments, be barely accessible.

***Vector field*** This post-processor plots the velocity field with vectors in different colors and lengths depending on the velocity.

***Streamlines*** A streamline is a tangent to the velocity field at a fixed time.

***Streaklines*** A streakline is also a tangent to the velocity field, but not at a fixed time. The streaklines follow the particles, as when a dye is injected into the flow. In steady-state flow, the stream- and streaklines are identical.

***Volume visualization*** Scalar variables such as pressure, electrical potential, and concentration may be visualized by using colors. Today's packages offer animation of intermediate files as well as export of reliable numerical data which can then be manipulated in external programs for further evaluation. Detailed visualization capabilities are one of the strong points of CFD techniques.

## 5.3
## Important Numerical Settings

Before undertaking a large simulation project it is crucial to have a simulation logbook. Even the most logical settings may be forgotten after a short while. The logbook should contain information about the grid generation, boundary conditions, and especially, the numerical parameters.

### 5.3.1
### Boundary Conditions

Boundary conditions are needed to define the problem and thus to obtain the solution. Surface boundary conditions are applied to so-called patches, as in CoventorWare$^{TM}$ [4]. A patch is a surface section of the model, e.g., a cube has 6 patches. In 2D problems a patch is a line. Here we briefly describe boundary conditions commonly used in EOF simulations such as EOF pumps [5].

***Pressure*** Specifies pressure on a patch.

***Voltage*** Specifies voltage on a patch.

***Inlet/outlet (default)*** Specifies zero-velocity gradients in the direction normal to a patch. One must be careful not to place the inlet/outlet in a nonuniform region, which would give spurious results.

**Symmetry**   Symmetry is invoked to save computational time. For example, two symmetry planes can reduce the number of cells by a factor of four. The symmetry BC assumes that gradients perpendicular to the symmetry plane or line are zero.

**Wall**   The wall BC simply sets the velocity on a patch to zero.

**Velocity**   Specification of a velocity on a patch. If needed, velocity profiles can also be specified.

**EOF mobility**   The commercial packages assume the Debye layer to be infinitely thin. This is valid if the Debye length is much smaller than typical channel dimensions. For an EOF pump, the typical Debye length is less than 10 nm, a factor a thousand or more smaller than the channel dimensions. Therefore, assigning EOF mobility to a patch is a good approximation. However, if the Debye length is comparable with channel dimensions, some corrected velocity boundary conditions must be applied [14].

### 5.3.2
### Solver Settings

When running software such as CFD-ACE+, which includes many numerical subroutines and parameters, the user must choose the proper setting to run the simulation optimally. Fine tuning is important because it affects convergence and can speed up solving the equations. If the parameters are not set and monitored properly, a simulation can yield wrong solutions that can appear correct. In the following, the most relevant solver settings are discussed.

*Numerical discretization schemes* are applied to discretize the partial differential equations in space and time, to compute a variable at a new position or time [12, 25]. However, the numerical discretization is not unique, and the following list points out strengths and weaknesses of the most commonly used schemes. Examples of *spatial discretization* schemes are:

- Central differencing schemes are applicable to diffusion-dominated low-Reynolds-number problems or when grid cells are small. They are not suitable for general flow problems involving convection, due to the fact that the direction of flow is not properly manifested. For high convection-to-diffusion ratios, these schemes yield oscillations or wiggles in the solution.
- Upwind differencing schemes are very stable and generally applicable, but have problems with numerical diffusion.
- Hybrid schemes are widely used, because they combine the central and upwind differencing schemes to take the advantage of both. However, they do produce some numerical diffusion.
- Higher-order schemes are more accurate, i.e., they minimize numerical diffusion, but are at the same time less stable, resulting in wiggles (so-called over- and undershoots).

*Temporal discretization* needs to be applied to time-dependent flows. Generally, smaller time steps are needed to achieve more accurate results. The proper time step needs to be determined for each specific problem, keeping in mind all relevant physical time scales. For discretized transport problems, the so-called Courant–Friedrich–Levy (CFL) number determines how many grid cells a fluid element passes during a time step [10]. Stability of simulations can be controlled by the CFL number [13]. For incompressible flows, the CFL number is defined by $CFL = v\,\Delta t/\Delta x$, where $v$ is the local velocity, $\Delta t$ is a time step, and $\Delta x$ a local cell size [13]. Knowing the velocity, the grid spacing, and the upper limit of CFL number (equal to or less than 1), one can get a feeling for correct length of the time step.

Three widely used *temporal* discretization schemes are:

- Explicit schemes that use only values of a variable from the previous time step to calculate the value at a new time. It is not generally recommended because it severely restricts the maximum allowable time step.
- The Crank–Nicolson scheme uses a combination (mixture) of the values at the previous and a new time step. It is usually used in conjunction with a spatial central differencing scheme. It has second-order accuracy.
- Implicit schemes use values from the surrounding nodes at the new time step. They have first-order accuracy. However, because it is unconditionally stable and robust, the implicit method is most recommended for general flows.

**Solver choice**  After the governing equations are integrated and discretized, they are converted into a system of algebraic equations. The matrix of this system is then solved by iterative methods assuring correct linkage between the pressure and the velocity fields. Two possible types of equation solvers are:

- The algebraic multi-grid (AMG) solver reaches the average solution quickly by using a hierarchy of grids, involving both fine and coarse grids, all the way down to a two-cell grid, to solve the equations. As opposed to a geometric multigrid solver, an algebraic solver manipulates matrices rather than the geometry. It is fast for both structured and fully unstructured grids. It is particularly good for low-Reynolds-number regimes where convection is low, such as in microfludics.
- The conjugate-gradient-squared (CGS) solver with preconditioning solves the system matrix in a straightforward way and is easier to implement, but at the expense of longer computational time. Matrix equations are treated as a minimization problem, and preconditioning is added to accelerate the convergence rate [15]. The solver is slower than AMG when a large number $N$ of grid cells is involved, because it takes $N$ steps to reach the end solution.

The default settings for these solvers rarely need to be changed. For thorough treatments of numerical solvers see [1, 15, 25].

**Residuals**  The governing equations are solved iteratively until convergence is achieved; in other words, when the solution no longer changes much from one iteration to the next. The residuals can be understood as differences between two

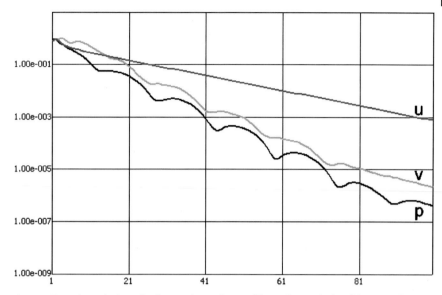

**Fig. 5.7** Typical residual graph of a simple 2D flow problem. The residuals of the two velocity components and the pressure are plotted on a log scale against the number of iterations. A drop of up to four orders in magnitude can be seen for all three variables. Calculated with CFD-ACE 6.6.

iterative values. The quality of convergence can be evaluated by monitoring the residuals during the iteration process (Fig. 5.7). The rate of change of a residual is usually plotted as function of iteration count. In a time-dependent simulation, the development of the residuals is presented at every time step. Generally, a calculation is considered converged if the residuals diminish by a factor of about $10^{-5}$, see Fig. 5.7. However, in some bubble simulations, when a special volume-of-fluid method for two-phase flows is used, convergence by more than two orders of magnitude is rarely achieved [1].

***Maximum iterations*** Although problem-dependent, the maximum number of iterations should be set to at least 40 or 50 and then adjusted according to the residuals.

***Relaxation parameter*** When it comes to practical simulations, setting the relaxation parameter causes the most frustration to users, so it is worthwhile to explain it in more detail. Depending on the value of the relaxation parameter, the terms over- and under-relaxation are used in literature, but we will use the shorter term 'relaxation' as it appears in the software.

The previously mentioned pressure–velocity linkage or coupling is somewhat tricky, and iterative algorithms need to be devised in a 'smart' way. For example, for incompressible viscous flow, there is no separate equation for pressure: pressure appears only via the gradient term in the momentum equations. The SIMPLE algorithm (semi-implicit method for pressure-linked equations) [16] reconstructs the pressure and velocity fields from an initial, guessed pressure. First, approximate velocities are calculated from the discretized momentum equations that include this initial pressure. Next, the continuity equation is used to obtain a pressure correction, which is then used to correct the pressure and velocities. The discretized equations for scalar variables such as temperature are then solved. The procedure is repeated until all variables converge [12].

Relaxation parameters are used to constrain the change in a variable from iteration to iteration, to prevent divergence. For example, the pressure correction might be too large, especially in the beginning when the initial pressure is far from its final value. Therefore, we want to relax the (non-relaxed) solution $\hat{S}_n$ at time step $n$, obtained by iterating the previous solution $S_{n-1}$. The relaxed solution $S_n$ is given by the expression $S_n = a\hat{S}_n + (1 - a)S_{n-1}$, where $a$ is a relaxation parameter between 0 and 1. Basically, all variables (velocity, pressure, temperature, etc.) are relaxed in the iterative process. Too-large values of $a$ cause large changes between iterations, leading to numerical instability (oscillations) and finally to divergence. However, too-small values of $a$ can cause extremely slow convergence [12]. Fine tuning of $a$ is important, because there is a minimum in the number of iterations needed to reach convergence for a specific value of the relaxation parameter [1].

The 6.6 version of CFD-ACE+ includes the specification of two relaxation parameters, which can be somewhat confusing. The above-defined parameter is called a *linear* relaxation parameter and is set to 1 by default. Decreasing this value adds stability and delays convergence. Because the program uses the SIMPLEC (SIMPLE consistent) algorithm [17], this value should be left unchanged and adjusted only if divergence is severe. For some skewed grids [18], the linear relaxation parameter for pressure was set to 0.8. (At the same time the pressure-correction relaxation parameter, introduced below, was set to the minimum value of $10^{-10}$.)

Usually, changing the *inertial* relaxation parameter has the opposite effect, i.e., decreasing it accelerates convergence at the cost of lower stability. It is important to recognize this difference, because users tend to keep increasing this parameter, which can actually take values greater than 1. One proven method is to modify it by systematically lowering its default value until convergence deteriorates or divergent behavior appears. Then one can be sure to be close to the above-mentioned minimum number of iterations. In many simulations, such as heat transfer, the inertial relaxation parameter must be decreased by several orders of magnitude. The code allows a minimum value of $10^{-10}$.

The relaxation parameter acts like a time step [25]. The elliptic equations, such as the Laplace equation for the potential, do not depend on time, and their variables can have the steepest change or largest 'time step' from one iteration to the next, i.e., the inertial relaxation parameter can be as small as possible (theoretically zero). As pointed out, the velocity is calculated from a guessed pressure that can be far from the final pressure; thus, the changes in both pressure and velocity between iterations need to be gradual to avoid divergence – a certain optimum relaxation value needs to be used. Generally, a larger inertial relaxation (smaller 'time step') is needed if convective terms are significant.

Here is a concrete summary of the inertial relaxation parameter for most common microfluidic variables (confirmed by experience):

- For velocity, values around 0.1 result both in fast convergence and good stability. In CFD-ACE+ the default value is 0.2.
- The pressure-correction inertial relaxation parameter is 0.2 by default, which is adequate for many simulations. Alternatively, it can be set to the smallest possible value ($10^{-10}$, corresponding to the largest 'time-step'), which is especially valid for low Reynolds numbers; if necessary, pressure can be regulated through the linear relaxation parameter as explained above. One of the reasons for using the smallest value for the pressure-correction parameter is that larger values prevent the algorithm from conserving mass at every iteration [19].
- For the electrical potential, the elliptic nature of the Laplace and Poisson equations allows the parameter to be the smallest possible to accelerate convergence.
- For species transport, in which convection is present, a certain relaxation parameter is necessary. The default value of 0.05 is usually sufficient.
- For heat-transfer simulations, the relaxation parameter for enthalpy should be as low as possible.

## 5.4
## Errors and Uncertainties

It would be useful now, after describing the various stages of the simulation process, to summarize the numerical errors and uncertainties involved. In CFD the following generalities can be made [13]:

*Model uncertainty*    real flow vs. exact solution of modeled equations.

*Discretization or numerical error*    exact solution vs. numerical solution of the equations (numerical diffusion enters this category).

*Iteration or convergence error*    fully converged vs. not fully converged solutions

*Round-off errors*    parameter values that are below machine accuracy.

*Application uncertainties*    lack of available data (precise geometry, specific BCs, diffusion coefficients, etc.)

*User errors*    mistakes and carelessness.

*Code errors*    bugs in the software.

## 5.5
## Interpretation and Evaluation of Simulations

Correct interpretation of numerical results is one of the most important issues, because of the complexity of simulations. In particular, the numerous visualizing capabilities of CFD programs, especially their nice display colors, can give a mis-leading sense of having obtained the correct result when this may not be true. So how do we know to what extent the simulation results represent reality?

The ultimate test of CFD results are experiments. In microfluidic systems it is rather difficult to make detailed measurements because the systems are small. However, one of the experimental techniques for flow measurements, namely par-ticle-image velocimetry (PIV), has been adapted for microfluidics, and simulations can indeed be directly tested against suitable experiments [20]. The idea behind the method is to reconstruct the flow velocity from the velocity of small fluores-cent particles (tracers) placed inside the flow. In simple language, the particles are illuminated by a light source and their positions are recorded in two consecutive instants by a CCD camera. From the known positions and the time delay between the two instants, the particle velocities and consequently the flow velocity can then be determined.

If experiments are not available, one powerful way is to compare the simulation with well-established theoretical models such as the equivalent circuit theory. This theory describes a fluidic network as an equivalent electrical network by expressing a linear relationship between a pressure drop (equivalent to a voltage drop) and the corresponding flow rate (equivalent to a current) [5, 21]. It is applicable to uniform incompressible flows and is very useful for microfluidic systems operating in low-Reynolds-number regimes. If the theoretical model agrees well with the simulation (in quantities such as pressure drop or flow rate), this is a strong indication that cor-rect results were obtained. Simulations, however, provide more detail than theory.

The accuracy of the algorithms in the software packages needs to be tested against experiment, analytical results, or some other form of approved reference values.

## 5.6
## Example Simulations

Both examples were simulated with CFD-ACE version 6.6.

### 5.6.1
### Fully-developed Flow in a Circular Capillary

In this example, the idea is to show several aspects of grid-dependency analysis. We consider Hagen–Poiseuille flow, i.e., pressure-driven flow in a circular capil-lary, for which an analytical solution exists. The validity of grids is determined by comparing the calculated velocities with the analytical solution.

*Physics*  When liquid from a large reservoir enters a straight tube of arbitrary but constant cross section, the velocity profile develops from being flat at the entrance to becoming paraboloid after a certain length, called the entrance length. For a cir-cular pipe, the entrance length can be up to 100 diameters [22]. Beyond the en-trance length, the velocity field is parallel to the cylinder axis, and its axial compo-

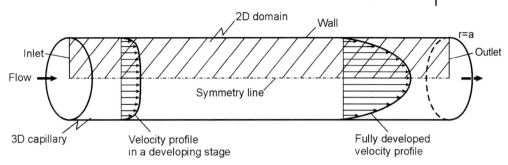

**Fig. 5.8** Schematic of a straight circular capillary of radius $a$. The dashed region is the reduced 2D computational domain with the assigned boundary conditions. In time-dependent simulations, a velocity profile develops through the entrance length to become paraboloid.

nent $u(r)$ depends only on the radial coordinate $r$ according to the well-known expression $u(r) = -(G/4\mu)(a^2 - r^2)$, where $G$ is the pressure gradient, $a$ the radius of the pipe, and $\mu$ the viscosity of the liquid. Here we consider fully-developed flows described by this equation. For microfluidic systems, which are connected to external reservoirs by long tubes, this is a good description [23].

**Geometry**  The problem is axially symmetric so the computational domain can be reduced to a 2D rectangle with four boundary conditions specified as inlet, outlet, wall, and symmetry, Fig. 5.8. The geometry of the computational domain is written as a Python-language script to facilitate changing the dimensions or the number of grid cells. Here, the following dimensions are used: $a = 75$ μm, $L = 900$ μm.

**Grid**  The velocity changes in the $r$ direction (orthogonal to the flow); thus, the grid resolution needs to be higher in that direction. Four different grids are shown in Fig. 5.9.

**Physical and numerical settings**  Because we are trying to solve a 3D case from the reduced 2D geometry, simulations need to be set to axisymmetric. The equations for axisymmetric problems differ from those for simple 2D problems: if a flow rate is specified at a 2D inlet, the solver integrates the value over the circular cross section to obtain the 3D value. The density and the viscosity of water need to be selected. The desired pressure is assigned at the inlet, and zero pressure at the outlet. Choosing the proper numerical parameters is governed largely by experience (section 5.3.2) and trial and error. Sometimes it is good to try the default values first and then make appropriate changes that yield the fastest convergence. The settings are summarized in Tab. 5.2.

**Simulation results**  The maximum velocities are calculated on each of the four grids. The program supplies the results in about five seconds. The velocity profiles from the coarsest and the finest grids are compared with the analytical solution in Fig. 5.10. The deviations between the analytical solution and the calculated

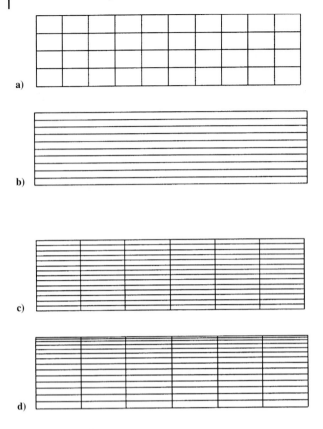

**Fig. 5.9** Four different grids of the 2D computational domain sketched in Fig. 5.8, with length $L = 900$ um and radius $a = 75$ um.
a) Coarse grid with only four cells in the vertical direction in which changes of velocity appear. The grid does not resolve velocity well enough (Fig. 5.10). There is no need for so many grid cells in the horizontal direction because there are no velocity gradients in the direction of flow.
b) Grid with better resolution in the vertical direction (10 cells) and only one cell in the direction of flow. This gives better results than the coarse grid (Tab. 5.3).
c) Fine grid with 14 cells in the vertical direction resolving the velocity profile very accurately (Fig. 5.10).
d) Biased grid with better resolution in the so-called boundary layer close to the wall where the gradients are the largest. It gives good results, although slightly less accurate for the maximum velocity calculated at the symmetry line. The biasing needs to be done close to the regions of the greatest changes.

values are summarized in Tab. 5.3. The most refined grid in the $y$ direction gives the best results, differing by only 0.5% from the analytical solution. It is important to notice the decrease in relative differences. Usually, no analytical solution exists for comparison, and analysis of the grid dependency must rely on the decrease in the deviations from one grid to the next-most-refined grid.

**Tab. 5.2** Numerical settings for the test case 1.

| Parameter | Value |
|---|---|
| Density | 0.9982 kg/m$^3$ (water at 20 °C) |
| Viscosity (dynamic) | 1.002×10$^{-3}$ kg/ms (water at 20 °C) |
| Inlet pressure | 100 Pa |
| Outlet pressure | 0 Pa |
| Temporal discretization scheme | None (steady state simulation) |
| Spatial discretization scheme | Upwind |
| Solver | Algebraic Multigrid (AMG) |
| Number of iterations | 100 |
| Inertial relaxation | |
| – Velocity | 0.2 |
| – Pressure correction | 10$^{-10}$ |
| Linear relaxation | |
| – Pressure | 1 |

**Tab. 5.3** Grid dependency analysis.

| Solution | $u_{max}$ (m/s) | Deviation in % (from analytical) | Deviation in % (from previous grid) |
|---|---|---|---|
| Analytical | 0.155938 | 0 | – |
| Grid 1 (coarse) | 0.150493 | 3.49 | 0 |
| Grid 2 (one cell) | 0.154811 | 0.72 | 2.78 |
| Grid 3 (fine) | 0.155089 | 0.54 | 0.18 |
| Grid 4 (biased) | 0.154967 | 0.62 | –0.08 |

The axial symmetry of the flow allows for a substantial reduction of the computational domain. If the flows are symmetric, but not axially symmetric, the simulations can be performed on half of the full computational domain, as in 5.3. When symmetry cannot be used, simulations need to be done on the full 3D domain. Although not necessary here, it would be instructive to highlight some grid aspects from simulations on a 3D Poiseuille problem, Fig. 5.11.

Fig. 5.12 shows three grids with different orthogonality properties used to mesh the complete circular cross section of the pipe. The orthogonality of grid cells directly affects the accuracy of flux calculations, and therefore the accuracy of results. For simulations of gas bubbles, only orthogonal grids yield convergence. For the above dimensions, the number of grid cells in the 3D domain is on the order of 6000.

All three grids reach the desired solution for the mean and maximum velocities within a 3% margin of error, although the third grid gives the best results. The deviation is acceptable but larger than in the above axisymmetric simulations. To obtain better agreement, the grids need to be even more refined, which significantly increases the simulation time. With the default solver settings, convergence in this 3D example was reached after 200 iterations in approximately 90 min.

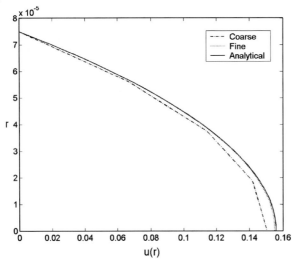

**Fig. 5.10** Theoretical and simulated velocity profiles. Note the four straight-line segments of the coarse grid, from each of the four cells.

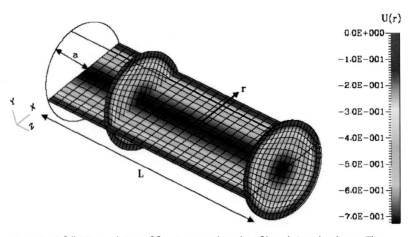

**Fig. 5.11** A full 3D simulation of flow in a circular tube of length L and radius *a*. The grids contain about 6000 cells, as opposed to 80 for the 2D axisymmetric example in Fig. 5.9. Courtesy M. J. Jensen.

5.6.2
**Movement of a Chemical Plug by Electroosmotic Flow in a Detection Cell**

In this example, the purpose is to show how numerical settings (time step, solvers, relaxation parameters) affect the results of time-dependent simulations. Both incorrect and correct results are presented.

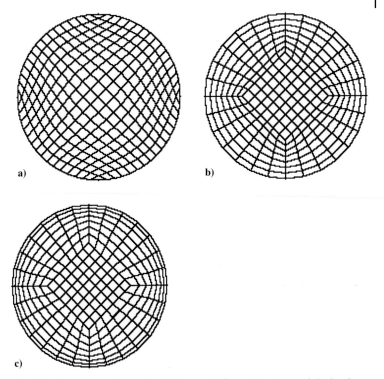

**Fig. 5.12** The three grids used in 3D Poiseuille flow. a) Regions with bad orthogonality at the edges. b) Good orthogonality at the edges and good aspect ratios. c) Good orthogonality and enhanced wall resolution. Courtesy M. J. Jensen.

**Physics** Fig. 5.13 shows a fluidic system used for separations [24]. By timely manipulation of voltages between sample, buffer, and waste reservoirs, a sample plug, consisting of several differently charged species, can be injected into the separation channel. The voltage drop generates an EOF, which is governed by the zeta potential at the walls. The EOF induces a downward bulk flow, and species from the plug, once they come into the separation channel, begin to separate due to different velocities in the applied electric field. When species enter the horizontal part of the detection cell, they are confined in the much narrower, longer region. If a light beam is then passed through this region, as from a waveguide fabricated next to it, the species absorbs the light. Due to the longer lightpath length, the detection signal in absorption measurements is significantly enhanced.

In making realistic simulations, the main concern is to correctly calculate the EOF, i.e., to couple the electric field with the flow. The potential distribution is found by solving the Laplace equation under voltage boundary conditions. The electric field at the boundaries is found as the negative gradient of the potential, and multiplying it by the value of the electroosmotic mobility yields the flow velocity at the boundaries. The mobility can be set as a BC or it can be calculated from the zeta potential. Heating phenomena are not taken into account here.

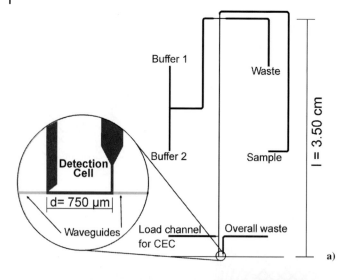

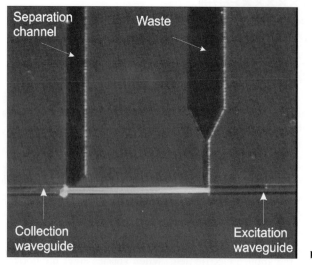

**Fig. 5.13** a) System of channels for a lab-on-chip used for chemical separations and absorption measurements [24]. The separation channel of length *L*=3.5 cm comes into the detection cell (enlarged) consisting of a long narrow horizontal part and two wave-guides. The geometry provides longer absorption length and thus enhanced detection. The widths of the separation and detection channels are 120 μm and 30 μm, respectively. The uniform depth of the channels is 12 μm. b) Photograph of the fabricated system. With permission from Wiley-VCH.

**Geometry** Simulations are performed on a 2D structure that includes only part of the detection cell, Fig. 5.14. This simplified setup reduces the computational domain but is sufficient to highlight the effects of (im)proper settings. However, 3D simulations that include the entire geometry with proper initial and boundary conditions

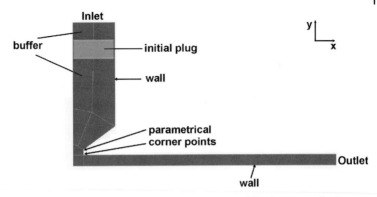

**Fig. 5.14** Computational domain representing part of the detection cell. The widths of the channels as well as the length of the horizontal channel are the same as in the real device, but the vertical part is only 400 μm. The geometry is 2D. Lines of the several structured domains can be seen; these can be separately meshed to achieve better control over grid resolution. Simulations performed with CFD-ACE 6.6.

are needed to perform qualitatively correct simulations comparable with experiments. Variables such as flow resistance are governed by the 3D geometry.

The real device is made of silicon, thus, its edges are sharp. The separation channel asymmetrically narrows to join the narrow detection channel. In this way dispersion of a plug entering the narrow part due to the race-track effect can be controlled. The two corner points that determine the junction between the wide and the narrow sections are parametrically specified for optimization purposes. The channel widths $w_{sep} = 120$ μm and $w_{det} = 30$ μm used in the simulations are the same as in the real device. In the CFD-ACE+ program, a special region within the geometry needs to be specified for a sample plug. The undistorted plug, consisting of a positive and a negative species, is positioned at the beginning between two lines parallel to the inlet (Fig. 5.14).

*Grid* Part of the grid is shown in Fig. 5.15. To make the grid orthogonal and well resolved around the corners, the geometry was divided into several segments. A larger number of cells in the separation channel produces smooth concentration profiles and allows better resolution of the concentration peaks. In the narrow region, the velocity profile is homogeneous and the overall (integrated) concentration is important, so the cell density can be reduced. However, biasing is implemented because the adjacent cells need to be of comparable size. The sharp corners produce singularities, leading to unreliable results in their vicinity.

***Physical and numerical settings*** Three modules need to be switched on at the same time: flow, chemistry, and electric.

***Fluid properties*** in this system, the solvent is water and the sample consists of positively charged rhodamine 110 ($C_{20}H_{15}ClN_2O_3$) and negatively charged fluores-

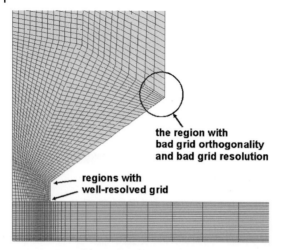

**Fig. 5.15** Part of the grid used in simulations. The cells in the middle of the vertical channel are fairly orthogonal. Orthogonality of the cells is not high close to the upper right corner. The areas around the two corner points in the narrow junction are well resolved. Biasing can be noted in the uniform horizontal channel. The layer close to the walls is not well resolved, especially in the right part of the vertical separation channel. This means that there is a large layer in which the velocity drops from a uniform EOF value (in the middle of the channel) to zero (at the walls). In reality, this is the Debye layer which is very thin compared to channel dimensions. In simulations it is not possible to completely resolve the layer; however, better resolution, e.g., with more cells and biasing close to the walls, yields a more accurate (smooth) velocity profile.

cein ($C_{20}H_{12}O_5$). In simulating various chemical solutions, the program uses the dilute approximation, which means that complicated interactions between and within chemical species are neglected. This is a reasonable assumption, because a sample is usually highly diluted in real experiments, with typical concentrations of $\sim 1$ nmol $L^{-1}$. Therefore, bulk properties such as viscosity, density, and conductivity of both the solvent and the sample can be assumed to be the same. To simplify things, the bulk properties of water were used. If, for example, a borate buffer $Na_2B_4O_7 \cdot 10H_2O$ was used, it would have the viscosity and density of water but a different conductivity, which depends on the buffer concentration and can be experimentally measured.

To assign specific properties such as charge, diffusion coefficient, concentration, etc., both the solvent and the sample need to be 'created' in the material database, so that they can be used with various settings, such as different initial conditions. For the sample species in our example, the diffusion coefficients were deduced from the diffusion constants of ions with similar structures [11]. The important properties are given in Tab. 5.4.

***BCs*** A voltage of 100 V is specified at the inlet and 0 V at the outlet. A zeta potential of –0.1 V was set at the walls. The boundary conditions are summarized in Tab. 5.4.

**Tab. 5.4** Property and BC settings for the test case 2.

| Parameter | Value |
|---|---|
| Density | $1 \text{ kg/m}^3$ (water) |
| Viscosity (kinematic) | $10^{-6} \text{ m}^2/\text{s}$ (water) |
| Conductivity | $10^{-4}$ S/m (typical value) |
| Diffusion coefficients | $10^{-9} \text{ m}^2/\text{s}$ (for both species) |
| Inlet voltage | 100 V |
| Outlet voltage | 0 V |
| Zeta-potential | –0.1 V |
| Debye thickness | $10^{-9}$ m |

**Tab. 5.5** Solver settings for the test case 2.

| Parameter | Improper | Proper |
|---|---|---|
| Time discretization scheme | Crank-Nicolson with default blending (0.6) | Crank-Nicolson with default blending (0.6) |
| Time step | 0.001 s | 0.001 s |
| Spatial discretization scheme | Upwind | Upwind |
| Solver | CGS + Preconditioning with default settings | AMG with default settings |
| Iterations per step | 25 | 30 |
| Inertial relaxation: | | |
| Velocity | 0.2 | 0.2 |
| Pressure correction | 0.2 | 0.2 |
| Voltage | 0.0001 | $10^{-10}$ |
| Species (concentration) | 0.05 | 0.05 |

**ICs** Initial conditions are particularly important in time-dependent simulations, because things change with each time step. The sample, selected from the material database, needs to be assigned to the plug region that was previously constructed with the appropriate geometry. The solvent needs to be assigned to the rest of the geometry.

**Solver settings** Two sets of simulations were performed. The 'bad' one included improper numerical parameters, yielding unphysical results; and the 'good' one with correct values. Both sets of settings are summarized in Tab. 5.5.

**Simulation results** The incorrect movements of the positive and negative species at different times are shown in Figs. 5.16a–d and 5.17a–d, respectively. The positive species seems to be moving too fast, so that the snapshots ($t$=0.10–0.15 s) barely resolve the concentration profile, a possible indication of a too-large time step. For the negative species, a clearly unphysical distortion of the plug occurs.

The movement of the species is a combination of three effects: the downward electroosmotic movement of the entire bulk fluid, i.e., the EOF, the electrophoretic motion of the charges toward the opposite electrodes, and diffusion. For negative species the upward electrophoretic motion opposes the EOF. We would expect the overall motion to be downwards, although less pronounced than for the positive species, because EOF usually prevails. From Fig. 5.17, it seems as though the EOF velocity is not well developed. A possible reason could be that the time step was set extremely small, which however is unlikely because the distortion is too pronounced at each time step. A more likely cause is that the potential did not converge fast enough. Because the potential delivers a physical force that drives the EOF, a converged potential is required to determine the proper velocities. Indeed, the results in Tab. 5.6 show that the residuals of the potential and the resulting velocity diminish by only two orders of magnitude after 25 iterations per time step. Unphysical recirculating regions in the velocity can be seen in Fig. 5.18a.

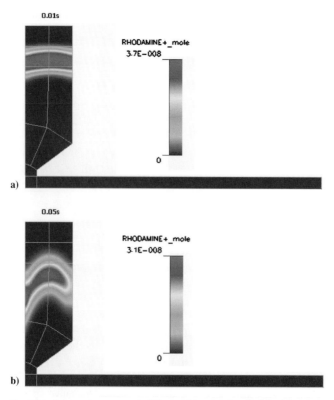

**Fig. 5.16** Incorrect movement of the positive species during electroosmotic flow. Concentration contours are shown, as well as four snap shots taken at 0.01 s (a), 0.05 s (b), 0.1 s (c), and 0.15 s (d). Unrealistic distortion at the edges can be noticed as well as a faded concentration profile that is unresolved in time.

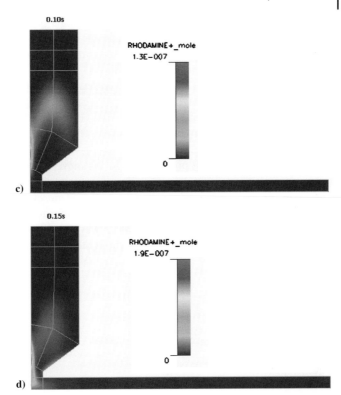

**Fig. 5.16  c, d**

**Tab. 5.6** Residuals of the electrical potential and the downward velocity in cases of improper and proper convergence. In the case of improper convergence, the electric potential drops by only three orders of magnitude.

| Iterations | Improper convergence | | Proper convergence | |
|---|---|---|---|---|
| | El. pot | y-velocity | El. pot | y-velocity |
| 1 | 0.1853 | 2.646 | 0.1152 | 1.708 |
| 2 | 0.005573 | 2.017 | 0.009248 | 1.41 |
| 3 | 0.003157 | 0.5955 | 0.001104 | 0.4346 |
| 4 | 0.001741 | 0.3222 | 0.000182 | 0.05882 |
| 5 | 0.001252 | 0.2257 | $5.69 \cdot 10^{-5}$ | 0.01116 |
| 6 | 0.001017 | 0.1873 | $2.24 \cdot 10^{-5}$ | 0.002832 |
| 7 | 0.00096 | 0.1568 | $8.40 \cdot 10^{-6}$ | 0.001407 |
| 8 | 0.000866 | 0.1344 | $4.15 \cdot 10^{-6}$ | 0.000612 |
| 9 | 0.000794 | 0.1169 | $1.74 \cdot 10^{-6}$ | 0.000336 |
| 10 | 0.000704 | 0.1033 | $8.62 \cdot 10^{-7}$ | 0.0002 |
| 11 | 0.000623 | 0.09193 | $3.97 \cdot 10^{-7}$ | 0.000132 |
| 12 | 0.000562 | 0.08224 | $1.94 \cdot 10^{-8}$ | $9.89 \cdot 10^{-5}$ |

The remedy is to increase the number of iterations per step or to choose a different (faster) solver. In addition, the relaxation parameter for the potential needs to be adjusted to speed up convergence. The correct time step was estimated from the condition CFL=1. (The size of a grid cell is $\sim 3\,\mu$m, the EOF velocity $v$ is $\sim 0.006$ m s$^{-1}$, yielding $t= \sim 0.0005$ s.) However, the step $t=0.001$ s from the first simulations was close enough, so it was not changed. Proper convergence of both the electric potential and the velocity can be seen in Tab. 5.6. The residuals drop ten and five orders of magnitude, respectively. The resulting, physically correct, velocity vectors are shown in Fig. 5.18b.

As a consequence of the proper settings, the movements of the species also appear physical (Figs. 5.19a–h and 5.20a–d). The positively charged rhodamine moves in an undistorted plug at the beginning. One of the advantages of EOF over pressure-driven flows is that it produces a uniform velocity profile, resulting in undistorted plug concentrations. That is why the distortions in Figs. 5.16 and 5.17 look suspicious. Even though the time step was not altered, it now takes more time for the species to enter the horizontal detection channel.

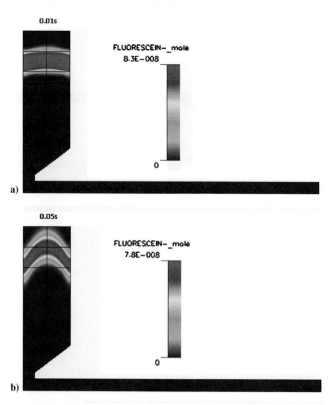

**Fig. 5.17** Incorrect movement of the negative species during electroosmotic flow. Highly distorted concentration profiles are indicative of a badly converged velocity, which in turn is linked to the electric potential (see Fig. 5.16).

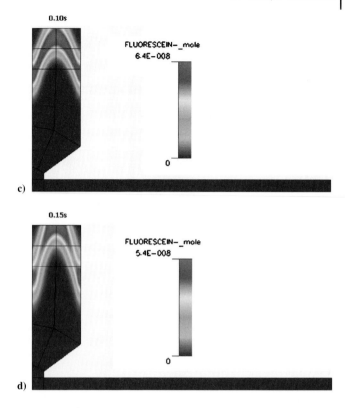

Fig. 5.17  c, d

More drastic changes appear in the movement of the negatively charged fluorescein: it moves only slightly downwards but in a completely undistorted way. After $t=0.35$ s, the two species are separated, with concentration peaks lagging behind each other, as can be seen by comparing the positions of the peak concentrations in Figs. 5.19 g and 5.20 d.

If the concentration profiles of, e.g., fluorescein from $t=0$ s to $t=0.35$ s are compared, we can notice an increase in the lighter, blurry region on both edges of the plug. This spreading of the concentration profile is caused by diffusion, whose magnitude is determined by the diffusion coefficient. Even if the diffusion coefficient is set to zero, some spreading occurs, due to the numerical diffusion implicit in some discretization schemes. We need to emphasize that simulations are reliable only within the certainty of the built-in models and the input data. Thus, an approximate diffusion coefficient additionally affects the accuracy of the output concentrations.

Corners represent discontinuities in boundary conditions and lead to singularities in numerical analyses. The results in their vicinity cannot be trusted. For a grid with a finite cell size, the singularity is smeared out over a region. The remedy is to make round corners or to choose different boundary conditions to represent the flow; for EOF the BCs can be velocity instead of the zeta potential. A

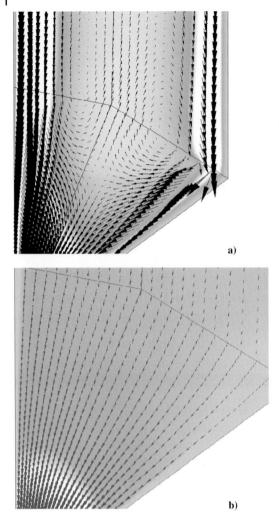

**Fig. 5.18** Velocity fields (vectors and contours). a) Unconverged velocity field with clearly unphysical recirculation areas caused by bad convergence of the electric potential (Tab. 5.6.).
b) Smooth, uniform velocity profile as a result of good convergence of the potential.

smeared region around a corner can be seen in the contour plot of the electrical field in the $y$ direction, Fig. 5.21.

***Evaluation*** As mentioned above, realistic 3D simulations are needed for appropriate comparison with experiment. Then the species concentrations integrated in the detection channel can be directly compared with the concentration-dependent

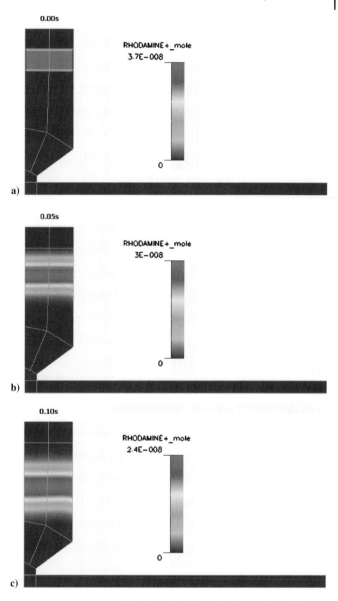

**Fig. 5.19** Correct movement of the positive species. The properly converged, uniform velocity results in an undistorted plug. It now requires more time for the plug to enter the detection channel. As the plug moves downwards, it diffuses both backwards and forwards. The two lighter regions broadening from the edges of the plug indicate the concentration spread due to diffusion. As indicated by the vertical concentration scale, the concentration maximum was reset at each time step, so the same colors represent different concentration values. This was done to highlight the shape of the moving plug, which would otherwise be barely visible.

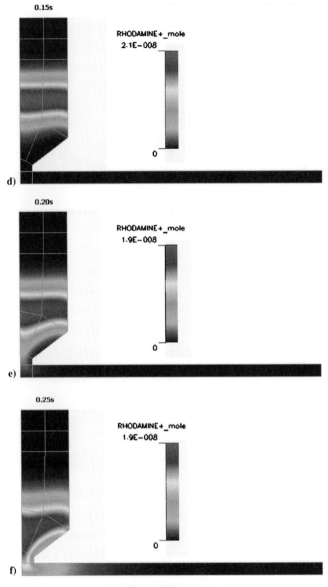

Fig. 5.19   d–f

signals from the absorbance measurements. Several things are worth noting when such a comparison is made.

Integration basically means summing the values of a variable from each grid cell of interest. The integrated signal depends somewhat on the grid resolution. Sharply discontinuous concentration peaks are usually a sign that the grid is not well resolved.

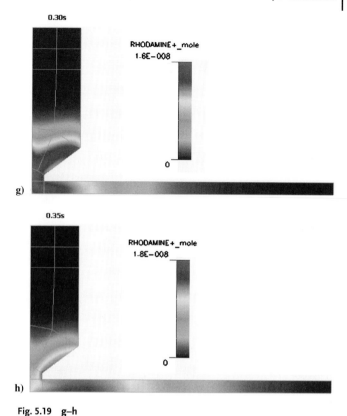

Fig. 5.19  g–h

CFD programs can display concentration values between certain specified limits. The lower limit for concentration needs to coincide with the limit of detection of the experimental setup. Therefore, only concentrations that can be detected should be displayed and integrated in simulations.

The effects of numerical diffusion need to be estimated.

Once the simulations reproduce the experimental results sufficiently closely, they can be used to optimize the geometries. In the above example, the slope of the asymmetrical narrowing can be optimized in relation to the sample distortion.

### 5.6.3
### Conclusions

The two examples should highlight some of the procedures used in setting up and interpreting simulation problems.

It is clear that using a CFD program requires physical and mathematical insight. The program itself may even contain a considerable amount of errors. Furthermore, special care should be taken when traditional numerical techniques are applied in new physical areas such as microfluidics. Therefore, it cannot be

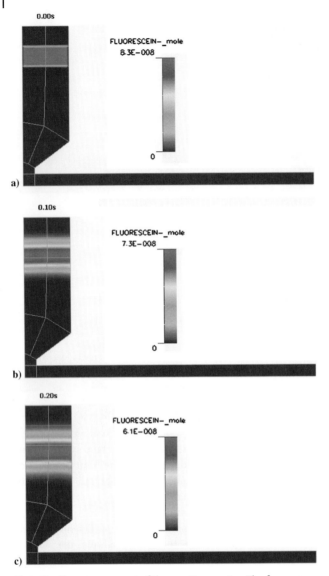

**Fig. 5.20** Correct movement of the negative species. The four consecutive snapshots indicate the expected downhill motion, which is less pronounced than for the positive species, due to the opposing electrophoretic effect. Comparing Figures 5.19 g and 5.20 d, we see that at $t=0.30$ s, the two plugs are completely separated.

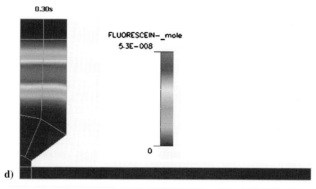

Fig. 5.20   d

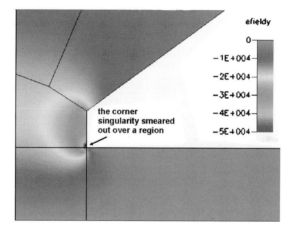

**Fig. 5.21** Contours of the γ-component of the electric field. The corner effects are visible, yielding large electric fields in the vicinity (dark spot), where the results cannot be trusted.

stressed enough how one should always weigh the results with a critical eye. Only through comparison with experiment or with proven theoretical models can simulations be truly confirmed.

**Acknowledgments**

We thank J. Maruszewski and Z. J. Chen of CFD Research Corporation for their input.

## 5.7
## References

1  CFD-ACE$^{TM}$ version 6.6, CFD Research Corporation, Huntsville, AL, USA, 2001; software, manuals and support: http://www.cfdrc.com/

2  JENSEN M. J., *Bubbles in Microchannels*, Masters Thesis, Technical University of Denmark, 2002; http://www.mic.dtu.dk/research/MIFTS/publications/msc.htm

**3** BITSCH L., *Blood Flow in Microchannels*, Masters Thesis, Technical University of Denmark, 2002; http://www.mic.dtu.dk/research/MIFTS/publications/msc.htm

**4** CoventorWare™ version 2001.3, Coventor, Cary, NC, USA, 2001; software, manuals and support http://www.coventor.com/

**5** BRASK A., *Principles of Electroosmotic Pumps*, Masters Thesis, Technical University of Denmark, 2002; http://www.mic.dtu.dk/research/MIFTS/publications/msc.htm

**6** HANSEN F. R., *Dispersion in Electrokinetically and Pressure-Driven Microflows*, Masters Thesis, Technical University of Denmark, 2002; http://www.mic.dtu.dk/research/MIFTS/publications/msc.htm

**7** MATLAB®, version 6, The Math Works, Englewood Cliffs, NJ, USA, 2000

**8** MATHEMATICA®, version 4.1, Wolfram Research, Champaign, IL, USA, 2002

**9** GORANOVIC, G., KLANK, H., WESTERGAARD, C., GESCHKE, O., TELLEMAN, P., KUTTER, J. P., Characterization of flows in laser-machined polymeric microchannels. In: *Micro Total Analysis Systems 2001*, RAMSEY, J. M., VAN DEN BERG, A. (Eds), Kluwer, Dordrecht, 2001

**10** MICHELSEN J. A., online CFD vocabulary, 1995; http://www.afm.dtu.dk/Staff/jam/vocabulary/vocabulary.html

**11** LIDE D. R., FREDERIKSE H. P. R. (Eds), *CRC Handbook of Chemistry and Physics*, 75th edition, CRC Press, Boca Raton, FL, USA, 1994

**12** VERSTEEG, H. K., MALALASEKERA, W., *An Introduction to Computational Fluid Dynamics, The Finite Volume Method*, Longman, London, 1995

**13** CASEY, M., WINTERGERSTE T. (Eds), *Best Practice Guidelines*, version 1.0, ERCOFTAC, 2000

**14** DUTTA, P., BESKOK A., Analytical solution of combined electroosmotic/pressure driven flows in two-dimension straight channels: finite Debye layer effects, *Anal. Chem.* **2001**, *73*, 1979–1986

**15** VESELY, F. J., Introduction to Computational Physics, online course, 2001; http://www.ap.univie.ac.at/users/ves/cp0102/dx/dx.html

**16** PATANKAR, S. V., SPALDING, D. B., A calculation procedure for heat, mass and momentum transfer in three-dimensional parabolic flows, *Int. J. Heat Mass Transfer*, **1972**, *15*, 1787

**17** VAN DOORMAAL, J. P., RAITHBY, G. D., Enhancements of the SIMPLE method for predicting incompressible fluid flows, *Numerical Heat Transfer*, **1984**, *7*, 147–163

**18** GORANOVIC, G., PERCH-NIELSEN, I. R., LARSEN U. D., WOLFF, A., KUTTER, J.P., TELLEMAN, P., Three-dimensional single step flow sheathing in micro cell sorters, in *Modeling and Simulation of Microsystems 2001*, LAUDON, M., ROMANOWICZ, B. (Eds), Computational Publications, Cambridge, MA, USA, 2001

**19** PATANKAR, S. V., *Numerical Heat Transfer and Fluid Flow*, McGraw-Hill, New York, NY, USA, 1980

**20** KLANK H., GORANOVIC, G., KUTTER J. P., GJELSTRUP H., MICHELSEN, J., WESTERGAARD C.H., PIV measurements in a microfluidic 3D-sheathing structure with three-dimensional flow behavior, *J. Micromech. Microeng.* **2002**, *12*, 862–869

**21** MORF, W. E., GUENAT, O. T., DE ROOIJ, N. F., Partial electroosmotic pumping in complex capillary systems, Part 1: Principles and general theoretical approach. *Sensors and Actuators B*, **2001**, *72*, 266–272

**22** WHITE, F. M., *Viscous Fluid Flow*, 2nd edition, McGraw-Hill, Singapore, 1991

**23** GRAVESEN, P., BRANEBJERG, J., JENSEN, O. S., Microfluidics – a review, *J. Micromech. Microeng.* **1993**, *3*, 168

**24** MOGENSEN, K. B., PETERSEN, N. J., HÜBNER, J., KUTTER, J. P., Monolithic integration of optical waveguides for absorbance measurements in microfabricated electrophoresis devices, *Electrophoresis*, **2001**, *22*, 3930

**25** PRESS, W. H., FLANNERY, B. P., TEUKOLSKY, S. A., VETTERLING W. T., *Numerical recipes*. In: The art of scientific computing, 2nd edition, Cambridge University Press, Cambridge, UK 1993

# 6
# Silicon and Cleanroom Processing

ANDERS MICHAEL JORGENSEN, and KLAUS BO MOGENSEN

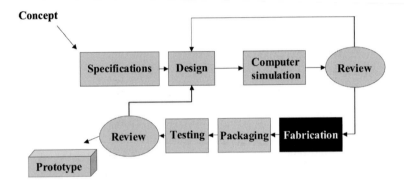

Silicon is now the best-known material and the one that can be obtained in the highest purity. Today pure silicon contains approximately one impurity atom (oxygen) per ten million silicon atoms, that is 99.99999% purity. The reason for the great interest in silicon is the semiconductor industry, where a continual increase in demand for components such as microprocessors and memory circuits (RAM) has driven research.

Many different techniques for manipulating silicon and silicon-compatible materials have been developed. They can be lumped together into a few general categories:

- substrate fabrication
- optical lithography
- deposition
- etching/removal
- heat treatment
- wafer level packaging

The following sections give an overview of the fabrication processes most often used for silicon-based lab-on-a-chip systems. The goal is to give an impression of the various processes and to give some hints on how the techniques can be applied.

*Microsystem Engineering of Lab-on-a-chip Devices*
O. Geschke, H. Klank, P. Telleman
Copyright © 2004 Wiley-VCH Verlag GmbH & Co. KGaA, Weinheim
ISBN: 3-527-30733-8

## 6.1
## Substrate Fabrication

Silicon substrates are used in very many of the devices fabricated so far for μTAS and probably will be used to a large extent in the future. The reason is that, up to now, silicon is the material that has been studied most intensively and that machining of silicon is at a very advanced state. Since the 1950s, the microelectronics industry has focussed mainly on silicon and, because the market for microelectronics products (microprocessors, memory circuits, etc.) is increasing exponentially, many resources have flowed into the development of more advanced strategies for manipulating silicon. This has resulted in silicon substrates being fairly cheap even with specifications that for other materials would be extreme, such as 1 foreign atom per $10^7$ host atoms.

The substrates are almost exclusively single-crystalline silicon. Silicon organizes into a crystal structure of the cubic type; it consists of two face-centered cubic cells, one of which is displaced a quarter of the lattice constant along the diagonal. This structure is called the diamond structure, because if is also the structure carbon takes when it forms diamond. Several key properties of silicon are given in Tab. 6.1.

The substrates used are disc-like wafers normally with a thickness of between 200 and 650 μm and diameters between 75 and 150 mm. For historical reasons the wafer sizes are often referred to in inches, even though metric measurements are actually used in specifications, for instance, a 100 mm wafer is also called a 4″ wafer. Today (year 2003) the 100-mm wafer size is the most often used size, but probably within 5–10 years, the standard size will be larger.

Wafers are usually cut to give top surfaces that roughly correspond to the main crystal planes in silicon given by the Miller indices (100), (110), and (111). For μTAS purposes, (100) wafers have been by far the most popular, because of the crystallographic-dependent wet etching possibilities. The wafers have parts of the disc cut to produce so-called flats. The primary or ordinary flat is cut in the [110] direction. It is used for coarse alignment of the wafer during photolithographic alignment. Flats can also be cut in other directions to provide information about the crystallographic direction of the wafer and also of the dopant type. The flats are usually cut according to the Semiconductor Equipment and Materials Interna-

**Tab. 6.1** Basic properties of single-crystal silicon.

| | |
|---|---|
| Density (g cm$^{-3}$) | 2.3 |
| Energy gap (eV) | 1.12 |
| Dielectric constant | 11.8 |
| Yield strength (GPa) | 3–7 |
| Thermal conductivity (Ω cm$^{-1}$ K$^{-1}$) | 1.57 |
| Thermal expansion coefficient ($10^{-6}$ K$^{-1}$) | 2.35 |
| Young's modulus (GPa) | 165 |
| Melting point (°C) | 1414 |

(100)  (111)

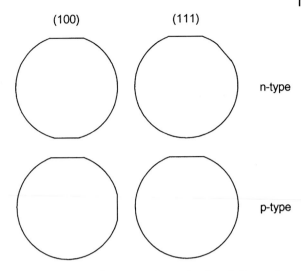

n-type

p-type

**Fig. 6.1** Position of primary and secondary flat of silicon wafers.

tional standard (SEMI standard). This means that the geometrical shape of a wafer reveals the crystal direction and polarity (Fig. 6.1).

Wafers are fabricated from ingots, which are large single-crystal rods. An ingot has a diameter somewhat larger than the desired wafer diameter and a length from 10 to 150 cm. These ingots are then ground to produce roughly the correct outside diameter and circular cross section. Then the crystal orientation is found by x-ray crystallographic methods or from anisotropic wet etching. Once the orientation is established, the primary flat is ground along the side of the whole ingot. Now the ingot is ready to be cut into wafers.

Saws are used to cut the wafers. Traditionally, rotating diamond-covered saws have been used, but a tendency towards wire saws is occurring. A rotating saw consists of a disc with a central hole somewhat larger than the ingot diameter. Along the rim of this hole is the actual saw blade, fitted with diamonds. The ingot is then placed in the hole and the saw cuts out the wafer. Silicon material is wasted where the saw cuts, especially because the saw blade must be fairly thick, $\sim 300\,\mu m$, to be rigid enough for cutting through the ingot. This large waste of otherwise valuable material has led to the development of the wire saw. Here, a thin diamond-coated wire about 160 μm in diameter is used. The wire is very long, 150–200 km, and is positioned so that many wafers can be cut at the same time. The wire saw saves material and is generally faster than the rotating saw; however, the rotating saw is more flexible in that wafers can be cut at angles other than the ingot direction. The rotating saw is also often used for low-volume fabrication, such as for microdevices, for which sometimes as few as 5 wafers are ordered at a time.

After cutting, the wafers are subjected to a lapping step. This is a coarse mechanical polishing using a slurry of aluminum oxide. Lapping gives a surface roughness of about 0.5 μm and removes about 50 μm of material from both sides of the wafer. Subsequently, one or both sides of the wafer are polished further, using finer mechanical polishing procedures. The surface roughness of a polished wafer side is typically 10–100 nm.

Ingots are grown using one of two methods: float zone (FZ) or Czochralski (CZ). For wafers that otherwise have the same specifications, FZ material is about 10–20% more expensive; however, the oxygen and carbon contamination of FZ material is lower than in CZ material. Both methods use a seed crystal as the starting point for the crystal growth; it is chosen to give the desired crystallographic direction to the ingot as a whole.

FZ material starts from a polysilicon rod of roughly the same size as the ingot to be produced. A seed crystal is placed near the end of the silicon rod and an electrical heater ring is placed near the tip of the polysilicon rod (Fig. 6.2). The diameter of the heater (RF coil) is a couple of centimeters. The heater begins the process by heating the polysilicon above its melting point (1700 °C), with the consequence that the molten silicon attaches to the seed crystal. The heater then moves up through the polysilicon rod, and in this way, the polysilicon is crystallized. At the same time as the heater moves, the seed crystal end rotates – the speed of rotation determines the diameter of the ingot. The FZ technique purifies the material as well as crystallizing it, because most of the materials that may contaminate silicon stay in the molten phase. This means that the crystalline silicon is purer than the polysilicon starting material.

If even purer silicon is needed, the resulting silicon rod can be recrystallized by the same FZ procedure. During crystallization, the ingot can be intentionally 'contaminated', or doped, by introducing dopants in the gas phase. Often, boron

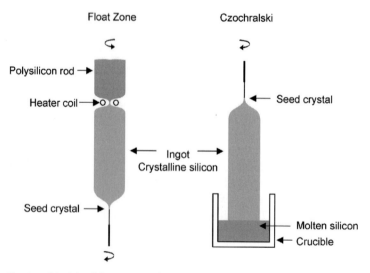

**Fig. 6.2**  Principle of float-zone and Czochralski crystallization methods.

**Tab. 6.2** Types of specifications for silicon wafers.

| Parameter | Comment |
|---|---|
| Wafer material | Silicon (single crystalline) |
| Growth method | Float zone (FZ) or Czochralski (CZ) |
| Diameter | Usually mm |
| Crystal direction | Usually Miller indices are used. Can include a range such as [100] ±1° |
| Primary flat | Usually Miller indices are used. Can include a range such as [011] ±1° |
| Secondary flat | Often stated as SEMI standard or Miller indices. Can include a range such as [01$\bar{1}$] ±5° |
| Resistivity | Given in $\Omega$ cm |
| Dopant type | $p$ or $n$ type |
| Dopant | Phosphorous (P) or antimony (As) for $n$-type |
| | Boron (B) for $p$-type |
| Thickness | Given in $\mu$m and includes a range such as (350±25 $\mu$m) |
| TTV | The difference in thickness between the smallest and the largest thickness of the wafer (best wafers have TTV <1–2 $\mu$m) |
| Wafer bow | The bending of the wafer (best wafers have bow <10 $\mu$m) |

doping is performed with diborane ($B_2H_6$) and phosphorous doping is done with phosphine ($PH_3$). Now (year 2003) the largest size of FZ material is 200 mm in diameter.

The Czochralski method uses a crucible in which silicon is melted. A seed crystal is then placed in contact with the molten silicon surface. The seed is pulled away from the melt and also rotated to give an ingot of the right diameter (Fig. 6.2). Dopants are introduced in the melt and incorporated into the crystal. Because of the crucible, CZ material is heavily contaminated with oxygen (up to $10^{17}$ cm$^{-3}$) and also has fairly high carbon content. This is often not a critical issue because the mechanical properties of silicon remain the same. If, however, electronic components are placed on the structure, it is often advantageous to remove the oxygen from the active areas. This can be done by special heating treatments and gettering methods. The Czochralski method can be used to produce very large ingots and today is the method used for fabricating wafers up to 300 mm in diameter.

Wafers are characterized by several parameters (Tab. 6.2). For different applications different parameters become important. For example, the total thickness variation (TTV) must be low if structures are to be fabricated with low tolerance for overetching. Normally the crystal directions are stated with an uncertainty, which are important for very precise crystallographic-dependent etches. A fairly simple wafer such as a 100-mm, single-side-polished (SSP), CZ, (100), SEMI standard flat, 500-$\mu$m-thick, $p$-type (boron), 1–30 $\Omega$ cm wafer costs about 8 € per piece if a couple hundred are bought at the same time. More specialized wafers may cost 50 € or even more.

## 6.2
## Optical Lithography

### 6.2.1
### Photolithography

An often-used technique for device fabrication is photolithographic masking. Here, a film of photoresist is applied to the substrate, and the photoresist is exposed to light through a photolithographic mask. After exposure, the photoresist is developed, which transfers the desired pattern to the photoresist (Fig. 6.3). When the substrate is subjected to a chemical treatment, the photoresist protects the surface, and thus the pattern on the mask is transferred to the substrate. The photoresist is removed by stripping, which is essentially dissolution of the photoresist in a nonuniform but fast way. Acetone is usually used to strip resist, but in some cases a special stripper must be used. The photoresist manufacturer generally makes such a stripper available.

Photoresists can be any of various photosensitive polymers. These polymers can be applied by different techniques such as spinning or spraying. In spinning, the thickness of the film can be expressed by this empirical expression:

$$t_{coating} \approx \frac{Kv}{\sqrt{\omega}} \qquad (6.1)$$

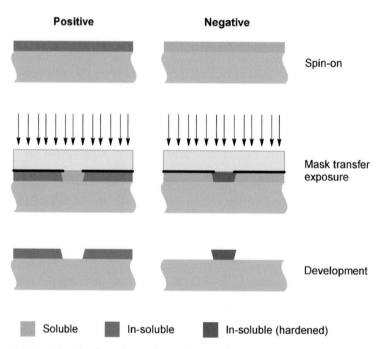

**Fig. 6.3** Principle of positive- and negative-tone photoresist.

Where $t_{coating}$ is the thickness of the coating, $K$ is a proportionality constant, $v$ is the kinematic viscosity (mm$^2$ s$^{-1}$), and $\omega$ is the number of revolutions per minute. The polymers are sensitive to light of particular wavelengths, and when exposed to these the chemical structure of the photoresist changes. Usually they are sensitive to UV light, with the i-line of mercury (365 nm) being particularly popular.

After exposure to light, the photoresist is developed by using a particular chemical solution. Some photoresists become more soluble in a developer after exposure, some become less soluble. The photoresists that become more soluble are called positive tone photoresists (Fig. 6.3); an example is AZ4562 photoresist. The other type of photoresist becomes less soluble after light exposure, due to cross linking of the polymer, and is called negative tone photoresist; an example is SU-8.

A third kind of photoresist is the so-called image-reversal photoresist (for example, AZ5214e). This type of resist is basically a positive photoresist, but after light exposure, the resist is subjected to an image reversal process by which the exposed areas become insoluble and the unexposed areas become soluble (Fig. 6.4). For AZ5214e, the image-reversal process is particularly simple, consisting of baking at elevated temperature (120 °C) followed by exposure of the photoresist without a mask, flood exposure. The advantage of image-reversal photoresists is that finer features can be fabricated than with other photoresists. It is also an advantage that the image-reversal photoresist allows one photoresist to be used for both tones; especially in a research laboratory it is advantageous to have this flexibility.

The size of the features that may be created with a particular photoresist depends on the light wavelength, the photoresist thickness, and the distance between the photoresist and mask. An approximate expression for the smallest mask feature, called the minimum linewidth, is:

$$w_{min} \approx \frac{3}{2}\sqrt{\lambda(s+z)} \tag{6.2}$$

Where $w_{min}$ is the minimum linewidth, $\lambda$ is the wavelength of the exposure light, $s$ is the separation distance between the photoresist and the mask, and $z$ is the thickness of the photoresist. Most often, the light source used for exposure of photoresist is a mercury spectral lamp and the i-line (365 nm) of the spectrum is used. When the mask is in physical contact with the photoresist coating ($s=0$), the alignment mode is called contact printing. But if there is some distance between the mask and coating ($s>0$), it is called proximity printing. According to the above equation, proximity printing gives lower possible resolution than contact printing; however, it puts less wear on the mask and there is no risk of damaging the photoresist coating on the wafer. Therefore, proximity printing is preferred for exposure of large, noncritical structures, and contact printing is used for critical structures and when tight linewidth control is desired.

When a pattern is transferred from a mask to a photoresist-coated wafer, a machine called an aligner is used. The aligner places the wafer and mask close together so that marks on the wafer and mask can be used for precise alignment of structures on the mask and wafer. The exposure time can usually be chosen so

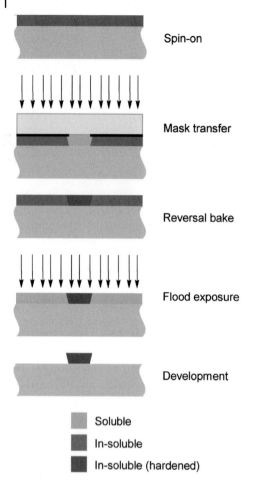

Spin-on

Mask transfer

Reversal bake

Flood exposure

Development

Soluble

In-soluble

In-soluble (hardened)

**Fig. 6.4** Principle of image-reversal photoresist.

that a timed shutter opens for a well-defined time. With some machines, the total exposure power is measured, which means that the total dose of light is controlled. The aligner is also responsible for handling the printing mode, for instance, making sure that the wafer and mask are parallel during proximity printing or controlling the force pressing the mask onto the wafer for contact printing.

A photolithographic mask is normally a glass plate with chromium patterns through which light cannot pass. A mask can be fabricated in several ways. One cost-effective approach is to use laser writing, which limits the linewidth to about 1.5 μm. The price is about 200 € per 125-mm mask and usually a delivery time of 1–3 weeks can be expected. If a finer linewidth is needed, electron beam-written masks can be used. These have a linewidth of approximately 200 nm; however, the price is as high as about 3000 € per 125-mm mask with an expected delivery time of several months. Further reductions in linewidth require different masking techniques usually employed only in the microelectronics industry; these techniques are very expensive and complex, usually requiring steppers or even direct electron-beam writing.

Cheaper and simpler solutions also exist: transparencies for overhead projectors are often used for simple, crude structures. A laser printer with a 600 dpi resolution creates dots of approximately 40 µm, and a 4800-dpi printer makes dots of 6 µm. Two to three dots are required for a reasonable pattern, so these systems can be used to give linewidths of 120 µm and 20 µm, respectively. The price for such masks is a few euros for the 600-dpi version and some 50 € for the 4800-dpi version. Masks based on transparencies suffer from several drawbacks: they are hard to clean and their dimensional stability is poor, due to warping of the flexible transparency. The problem with dimensional stability has considerable impact on the alignment precision. Alignment of transparency masks becomes very difficult if the precision needed is better than $\sim 200$ µm (on a 100-mm wafer).

The drawbacks of masks based on transparencies can be overcome by transferring the pattern from an overhead mask to a glass substrate covered with chromium and photoresist. After development and etching of the chromium, the formed mask can be used just like a laser-written or electron-beam mask and can be cleaned by the same cleaning methods.

Consider a structure such as a channel network with metal electrodes. This structure requires two different patterns: one for the channel network and another for the metal electrodes. These patterns are placed on two different photolithographic masks. First one mask is used, together with photoresist, and later the other mask is used. The desired structures, however, require the second mask to be aligned within a certain tolerance with respect to the first mask. This tolerance is often on the order of a few micrometer. An aligner that can hold both the mask and the substrate and allow them to move with respect to each other is used. The aligner typically also contains the UV light source used for exposing the photoresist.

## 6.2.2
## Mask Design

To make photolithographic masks, a mask design is created. This is almost exclusively done with computer-aided design (CAD) software such as L-Edit<sup>tm</sup> (Tanner Research, USA). The different masks appear as different layers or colors in the layout (Fig. 6.5). It is important that the software can handle various zoom ranges, because a structure often includes patterns on the order of microns and the wafer itself is 100 mm or larger. Several standard computer-file formats are used when creating masks: Caltech intermediate format (CIF) and Calma stream (GDSII).

The CIF format is a text file that can be read and edited by most text editors. It has a few simple commands such as: box (drawing rectangles), dot (drawing circles), and poly (for drawing polygons). It also supports reuse of patterns, through so-called symbols. A CIF symbol is a pattern, which can contain patterns on some or all of the layers in the design. The pattern described by the symbol can be of any useful size. The symbol can then be instanced (inserted) at several places in the design,

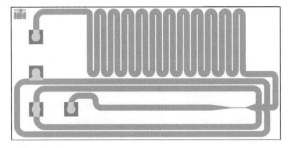

**Fig. 6.5** Example of a chip made from a mask design. The four squares on the left side are in one mask layer, and the meandering channel is in another. Where the two layers overlap, the overlapping structure is shown as a lighter color. The chip size is 10 mm by 20 mm.

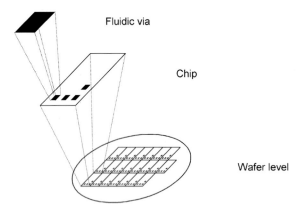

**Fig. 6.6** Sketch showing the use of symbols. The fluidic access point, drawn as a simple square, is defined as one symbol. This symbol is then repeated 4 times on the chip, which itself is defined as another symbol. Finally, the chip symbol is instanced 22 times to create the full wafer-level layout. Changing the fluidic access points on the whole mask is thus a matter of changing just one symbol.

while being defined only once. The use of symbols is encouraged, because it helps to keep the design consistent and easy to update and also decreases the file size

Consider for example a structure in which the chip size is 10 mm by 20 mm (Fig. 6.5): a 100-mm wafer can contain 22 chips. Each chip has four fluidic access points. By using symbols, one could create a fluidic access point as one symbol and instance it 4 times in a chip. Then, the chip is used as a symbol and instanced 22 times to create the final layout (Fig. 6.6).

Compare this to the situation where the drawn structures are merely copied. The information for drawing 4 fluidic access points on each of 22 chips is stored as 88 items. What if the pattern for the fluidic access has to be changed? With the symbol approach, only one symbol is changed, and its instances throughout the layout are automatically updated. However, with the copying approach, all the individual patterns have to be deleted and new patterns must be copied into the correct positions.

In general, it is advantageous to keep the number of actual photomasks low, because the whole photolithographic cycle is time consuming. On the other hand, sometimes more masks should be used, simply to make processing much easier.

**Alignment marks** Sometimes it can be a great advantage to use a mask to define some alignment marks for subsequent masks, for example, when the alignment-mark mask allows more freedom in the order of using the subsequent masks. It may also happen that subsequent masks cannot include alignment structures, because processing these layers destroys any fine markings they may contain (for instance, deep isotropic etches). When designing alignment marks, one must keep in mind that they should not be vulnerable to processing, but must be easy to find, easy to use, and hard to misinterpret. Two sets of alignment marks are used, one on the left side of the design and another on the right side. For highest precision, the distance between alignment marks should be as large as practically possible.

On some aligners, the field of view may be as small as 100 by 100 μm, so, to ensure that alignment marks are easy to find, it is necessary to have distinct helping structures, such as arrows and pointers, to guide the aligner operator to the alignment marks (Fig. 6.7). Designing alignment marks for ease of use is a matter of experience and personal preference. Fig. 6.7 shows an example of a reasonably easy to use structure. Each of the 5 alignment structures consists of 7 bars surrounding 2 squares of unequal size. The bars are on the alignment mask and the squares are on the mask to be aligned to the alignment mask. The 5 structures are different in that the central bar has a different length from its neighbors. The

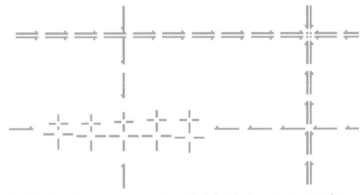

**Fig. 6.7** Part of an alignment structure. The left side shows 5 structures for aligning other masks to this one. The right side shows a structure for aligning the front to the back. The width of each line is 10 μm.

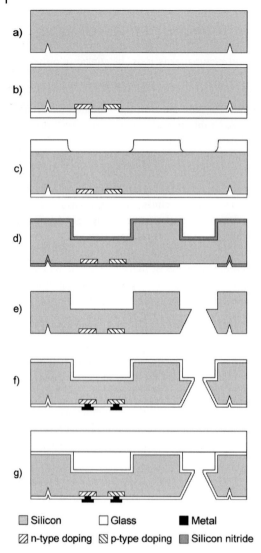

a)

b)

c)

d)

e)

f)

g)

**Fig. 6.8** Example of fabrication sketch showing a cross-sectional view of a device for some of the steps required to fabricate an integrated chemiluminescence detector for use with enzymes.

☐ Silicon     ☐ Glass     ■ Metal
▨ n-type doping   ▧ p-type doping   ▨ Silicon nitride

pattern shown is the left-side structure; a mirror version is placed on the right side. The configuration with mirrored versions and two unequal squares as the alignment object means that it is immediately obvious if the wrong mark is used or if the mask or wafer is placed upside down in the aligner. Preventing such simple handling mistakes is relatively easy if prevention is designed in, but a lot harder to do later on. Another example is to include a large logo or other structure on the masks, which makes it immediately obvious how the masks are supposed to be placed with respect to each other.

## 6.2.3
### Hints in Planning Fabrication Runs

Designing the masks and the fabrication sequences needed for creating a particular lab-on-a-chip requires some thought and overview. Usually there is no set fabrication sequence to be used, which is contrary to the microelectronics field, where many specific standard processing sequences exist (e.g., MOSSIS). For lab-on-a-chip and most micromachining purposes, the sequence and actual processes can be chosen more or less for the specific system at hand. This enormous flexibility may be daunting for newcomers to the field.

One approach to dealing with this is to use drawings and figures to show what an individual step is supposed to do (Fig. 6.8) and then to find the process needed to do just that.

Creating several different sketches of each device helps the designer to clarify which processes may be replaced with others and also to find out which processes are critical for successful fabrication. Either simultaneously with the sketches or later, the masks can be designed. The level of detail employed when creating the sketches is a matter of personal preference; however, it is best if the level is detailed enough that others can understand the sketch and description. One should also carefully consider what happens on parts of the design other than the actual device area, because processing can jeopardize the structural integrity of the wafer, causing it to break or have holes unless special care is taken.

## 6.3
### Deposition

This section describes, first, some fundamental properties of thin-film coatings; second, several deposition techniques, focusing on the underlying processes; and third, several specific materials, including how they are deposited, what their properties are, and where they are used.

### 6.3.1
### Fundamentals of Coatings

Several concepts are used in describing a thin-film coating, the most important ones are uniformity, homogeneity, conformity, adhesion, and stress.

*Uniformity* describes by how much the thin-film thickness varies across a wafer. Uniformity can be expressed precisely as a variation range, such as uniform within 10% or within 100 nm. This is the variation between the maximum and minimum measured thicknesses. It can also be expressed more qualitatively by stating that the film is uniform or nonuniform. These qualitative expressions are naturally hard to gauge and should be avoided when describing specific films.

*Conformity* describes how well the coating can cover steps in the surface. A conformal coating covers every surface to the same thickness. Conformity is a local description, which means that a coating can be conformal, yet still be nonuniform. There are several ways in which a film can be nonconformal. Some films might not cover sidewalls, whereas others may have protruding corners overhanging the bottom of steps. Some deposition methods may yield nonconformal coatings simply because no material can reach into the depth of the step.

*Homogeneity* describes the composition of the film. For instance, if the concentration of a certain compound varies throughout the film, then the film is nonhomogeneous. The term homogeneity is almost always used qualitatively because, in general, any film has variations in concentration if the length scale is small enough. That is, on an atomic scale all films show variations in composition. Films with density variations can also be termed nonhomogeneous.

*Adhesion* describes how well the film sticks to the surface and can be expressed as the energy needed to pull the film off the surface. It can also be stated in more qualitative terms. Sometimes a film sticks so well to the surface that attempts to pull the film off result in the substrate material being ripped apart instead. Such a situation is called bulk fracture and means that the physicochemical binding between atoms on the surfaces of the film and the substrate is stronger than the binding between the atoms comprising the substrate.

Many of the processes used during microfabrication introduce stress in the substrate. Stress is a term used to describe certain mechanical interactions between the film and the substrate. If a film is applied and causes the surface of the substrate to bend upwards, the film possesses tensile stress (Fig. 6.9 left). But if the film causes the substrate to bend downwards, the film possesses compressive stress (Fig. 6.9 right).

Consider wafer bending due to tensile stress. At the time of deposition, the wafer is flat. As a result of tensile stress, the film contracts, relieving part of the tensile stress, but at the same time slightly compressing the underlying wafer. Now both the film and the wafer are in force equilibrium. The film is still stretched, although less than before, and the wafer is now compressed. Judging from the force equilibrium, the film pushes the wafer together and the wafer pulls the film apart.

Stress is measured in units of pressure, as pascals or dynes cm$^{-1}$. The value is independent of the thickness of the film. The extent of substrate bending, however, is highly dependent on film thickness. When a wafer is subjected to a stressed film it forms into a bowl-like shape. The radius of curvature can be used, together with some material data for the film and substrate, to find a numerical value for the stress with the Stoney equation:

**Fig. 6.9** Wafer bending due to stress. Left, bending due to tensile stress; right, bending due to compressive stress. The film is the thin coating on top of the wafer.

$$\sigma = \frac{1}{R} \frac{E}{6(1-v)} \frac{T^2}{t} \tag{6.3}$$

where $\sigma$ is the stress, $R$ the radius of curvature, $E$ Young's modulus of the substrate, $v$ the Poisson ratio, $T$ the substrate thickness, and $t$ the film thickness. The sign is negative for tensile stress and positive for compressive stress. However, to avoid confusion it is usually clearer to state whether the stress is tensile or compressive instead of depending on the sign.

In many applications, a thin-film coating is used as a membrane; the film is applied first and then a cavity is formed underneath it. If the film possesses tensile stress the membrane is flat and strained. If the film has compressive stress, it might bend and buckle.

The stress level of a thin film depends on the material, the deposition parameters, and the maximum temperature the coating has been subjected to after been placed on the wafer. Differences in the thermal coefficient of expansion between the film and the wafer contribute to stress.

### 6.3.2
### Deposition Methods

*Chemical vapor deposition*  One approach to depositing material is chemical vapor deposition (CVD). The wafers are exposed to some chemical compounds in gas form. If the conditions are right, the chemicals react on the surface and a coating is created (Fig. 6.10). During the deposition, reactive products are formed in the gas phase and are transported to the surface where they are adsorbed. They can move around on the surface while adsorbed. This movement is called surface diffusion or migration, and the distance moved is called the migration length. The reactive products can then take part in a surface reaction, reacting with other

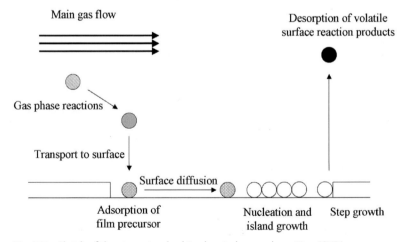

**Fig. 6.10**  Sketch of the stages involved in chemical vapor deposition (CVD).

migrating compounds (forming islands) or with the surface (step growth). Afterwards the byproducts must desorb and be transported away.

The growth rate and uniformity of the film depend strongly on the supply of reactants, removal of byproducts, and the reaction rate on the surface. The conformity of the film depends strongly on surface migration. Conformal coating is obtained when the migration length is long, and nonconformal coatings arise when the migration length is short. The migration length is usually longer at higher temperatures and shorter for high surface densities of reactants or very high surface reaction rates. Two of the most popular CVD methods are low pressure CVD (LPCVD) and plasma-enhanced CVD (PECVD).

Low-pressure chemical vapor deposition usually takes place at about 100–300 mTorr and elevated temperatures. The LPCVD reactor resembles a furnace, and a batch of wafers is usually loaded. LPCVD reactors are usually operated so as to have a long surface migration distance and excess chemicals in the gas phase. This generally results in high-quality, uniform, homogeneous, conformal films of excellent properties. However, the downside of these operating conditions is high temperature and slow growth.

Plasma-enhanced chemical vapor deposition uses plasma to dissociate the reaction molecules, which then generally become highly reactive. Together with a fairly low temperature (300–400 °C), this leads to short migration lengths, which in turn lead to nonconformal coatings. PECVD is very popular because of the low temperature, which allows coatings to be made on metal lines. It is also popular because of generally high deposition rates and high flexibility in operation conditions. The composition of films can be changed by adding dopant gases. For instance, silicon dioxide deposited in a PECVD reactor can be doped with phosphorous, boron, nitrogen, germanium, aluminum, rare earth metals, etc. The stress level, hydrogen content, and porosity are generally also affected by the composition and the physical conditions under which deposition occurs.

A common PECVD setup uses two parallel plate electrodes and radio frequency electrical excitation. The wafers are placed near the bottom electrode (cathode) and electrical power is supplied to the electrodes, normally at a frequency of 380 kHz or 13.56 MHz at up to several kilowatts of power. Reaction gases are then fed into the chamber through the side of the chamber or through the top electrode. If the gases are supplied through the top electrode it is usually formed as a showerhead with multiple small openings to facilitate even gas coverage.

A parallel plate reactor resembles the sputtering setup, described below, in many ways. This means that the positive ions in the plasma are accelerated towards the surface and, if they reach sufficient energy, might remove parts of the surface. This so-called sputtering effect competes with the deposition effect; it increases at higher power and lower pressure and decreases at higher plasma density.

***Sputtering*** In sputtering, positive ions from a plasma bombard a target of the film material. The bombardment knocks atoms from the target, which then move into the plasma as neutral species. These atoms collide with the plasma atoms, which results in atoms moving in all possible directions. If the wafer to be deposited is close to the plasma where the target atoms move, the wafer receives coating atoms from all angles, and the coating can be fairly conformal. But if the wafer is placed far from the plasma, the only atoms able to reach the wafer have directions that are fairly close to each other.

The ions need sufficient energy to knock out the target atoms to achieve appreciable sputtering. An electric field is used to accelerate the ions onto the target, by keeping it at a low potential (cathode). To increase the efficiency of sputtering a magnetic field is often introduced near the target. The combination of electric and magnetic fields traps electrons near the surface and thus increases the interaction between electrons and target. Sputtering involving a magnetic field is called magnetron sputtering and is the most popular approach.

One reason for the widespread use of sputtering is the ability to deposit alloys and materials with a high melting point, such as tungsten. The plasma used for sputtering is sometimes used as part of the coating. This most often involves using oxygen as plasma and then depositing the oxidized form of the target material as the thin-film coating.

***Resistive heating deposition*** One approach to creating a thin-film coating is to create a cloud of vapor of the desired material and then present the substrate to this cloud. This can be done by heating a crucible containing the coating material. The material must be above its melting point before any useful amount of matter enters the vapor phase. Generally, such heated material sticks to and coats any cold surface. It is important to have vacuum conditions when depositing in this manner, because the atoms in the vapor should hit the wafer before being cooled by any atmosphere in the chamber. If the pressure is too high, then there are too many foreign atoms in the deposition chamber. Under these conditions, the material in the vapor might condense and form nano- or microparticles. Such particles make the film nonuniform and less homogeneous.

***Electron-beam deposition*** Another approach to creating a vapor of a material is to heat it locally, e.g., by bombarding the material to be deposited with a beam of electrons. The energy carried by the electrons is transferred to the material, heating it. This approach requires high voltages ($\sim$kV) and for the same reasons as for resistive heating, deposition requires fairly high vacuum conditions. Electron-beam deposition is used often, because it is fast and can be used for a wide range of metals, requiring only small amounts of the film material. Another advantage is that the material reaches the wafer at a specified angle.

Usually the material approaches perpendicular to the surface. This results in a relatively nonconformal coating that can be exploited for the lift-off method of patterning. E-beam deposition can be used for depositing layers ranging from 2 nm to several micrometers in thickness, depending on the time used and power fed into the target.

***Spin-on coatings*** If liquids are used as a coating material, spinning is often used. The thickness of the film depends primarily on the viscosity of the liquid and the maximum spin speed. The conformity and uniformity of the films depends on the details of the spinning procedure. Such procedures may involve many steps affecting spreading, flowing, thinning, and rest parameters to achieve the desired film properties. Normally a spin-coating step is followed by a heat treatment in which solvent is evaporated and/or the film is solidified. Most materials that can be spun on can also in theory be sprayed on by equipment resembling airbrushes. However, several technical complications are introduced by spraying. For instance, an aerosol spraying can produce charge buildup, which can lead to sparking. Because the solvents used are often highly flammable, several measures must be taken to avoid explosions. This has so far meant that the price for dedicated spraying equipment is very high and that spin technology has advanced farther than spray technology.

***Electroplating*** Electroplating can be used to form very thick layers of metals or alloys, for instance, in creating thick mechanical structures out of metal or in creating tools for injection molding. One example of deposition works by having a metal salt in solution; the wafer to be coated is placed on one electrode (cathode). The other electrode (anode) is immersed in the electrolyte (Fig. 6.11). By controlling the potential of the two electrodes a current is forced through the system.

At the cathode, the metal ion accepts a certain number of electrons and is deposited on the wafer; on the anode another reaction releases electrons. An example of nickel deposition using nickel chloride is:

$$\text{Cathode}: \quad Ni^{2+} + 2\,e^- \quad \rightarrow \quad Ni$$

$$\text{Anode}: \quad 2\,Cl^- \quad \rightarrow \quad Cl_2 + 2\,e^-$$

The current passing through the system controls the deposition rate, and comparatively high deposition rates can be achieved (several tens of micrometers per

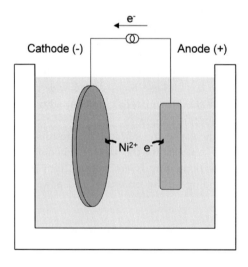

**Fig. 6.11** Principle of electroplating.

hour). Because of inhomogeneous electric field distribution, electroplating often results in a relatively thick edge layer unless special precautions are taken. It is crucial for this type of deposition that there is electrical contact to all the areas that are to be metallized.

A variation of electroplating is electroless plating, which is a method that does not require an external electron source. Instead, the solution is unstable and metal is deposited on any surface that catalyzes the reaction. Electroless deposition is generally slower than electroplating, and the chemistry involved is much more complex. This means that in general electroplating is preferred over electroless plating.

### 6.3.3
### Materials

*Silicon*  Silicon can be deposited by various means. Usually the deposition processes results in amorphous silicon ($\alpha$Si) or polysilicon (pSi). Polysilicon possesses short-range order, so that polysilicon is formed of grains within which the diamond structure of crystalline silicon exists. These grains are then randomly organized with respect to their neighbors so that no long-range order is present. Polysilicon can be deposited by LPCVD by this reaction:

$$SiH_4 \quad \rightarrow \quad Si + 2\,H_2$$

Polysilicon is deposited at a deposition rate of about 5 nm min$^{-1}$. If the temperature of the reaction is decreased, the grains become smaller. When the grains become very small and effectively cease to exist, the material is amorphous and possesses no order on any range. This material, called amorphous silicon, can also be deposited by sputtering of a silicon target, where the target is usually *p*-type doped to be conductive.

A third way to deposit $\alpha$Si is PECVD. This method is also based on silane (as for LPCVD); however, the films contain substantial amounts of hydrogen (up to 20%) and are sometimes called hydrogenated amorphous silicon films, $\alpha$-Si:H. These films have superior electrical qualities compared to polysilicon and other amorphous silicon films. Both LPCVD and PECVD silicon films can be doped in-situ by boron from a diborane ($B_2H_6$) source or by phosphorous from phosphine ($PH_3$). Alternatively, the films can be doped as silicon using predeposition, gas phase, or implantation methods.

*Silicon nitride*  Silicon nitride is an important material because it can function as a mask for KOH etching, is highly durable, and normally has tensile stress. Silicon nitride in the stoichiometric form $Si_3N_4$ has a stress level of approximately 2 GPa, and can cause surface rupture on silicon wafers (along the [100] or [110] directions) for layers of 700 nm. Thicker layers tend to crack in the film or at the surface. Stoichiometric silicon nitride has a refractive index of 2.0 (at 632 nm) and a bandgap of about 5 V. Stoichiometric silicon nitride can be made by LPCVD by this reaction:

$$3\,SiCl_2H_2 + 4\,NH_3 \quad \rightarrow \quad Si_3N_4 + 6\,HCl + 6\,H_2$$

Nonstoichiometric silicon nitride is often used, because it has many of the same qualities as stoichiometric silicon nitride, such as chemical resistance and durability. However, the nonstoichiometic nitride contains a larger fraction of silicon, which decreases the stress in the film. Depending on processing parameters and heat treatments, films with zero total stress can be obtained. This means that fairly thick films can be deposited, with thickness of around 3 µm commonly used for membranes.

Nonstoichiometric silicon nitride is given various names such as SiN, $SiN_x$, low-stress SiN, and low-$\delta$ silicon nitride. None of these names are accurate descriptions of the material and further information must be given to know exactly what material is involved. Nonstoichiometric silicon nitride contains more silicon than stoichiometric silicon nitride, which can be observed as an increase in refractive index (by up to 2.4) and a decrease of the bandgap. The bandgap reduction means that SiN absorbs light at longer wavelengths than does stoichiometric silicon nitride.

Silicon nitride films can also be deposited by PECVD. The deposition follows this reaction:

$$SiH_4 + NH_3 \quad \rightarrow \quad SiNH + 3\,H_2$$

The films are hydrogenated, which means that the refractive index is decreased. The film might also be porous, additionally decreasing the index. Alternatively, it might also contain more silicon, which raises the index. Thus, not much information is gained from knowing the refractive index.

***Silicon dioxide***   Silicon dioxide can be deposited by many different methods (see also thermal oxidation, section 6.6.1). Conformal LPCVD coatings can be accomplished by several different reactions. One of the most important is the use of tetraethylorthosilicate, $Si(OC_2H_5)_4$, or TEOS. This compound decomposes directly to silicon dioxide:

$$Si(OC_2H_5) \quad \rightarrow \quad SiO_2 + products$$

The deposition rate is about $10\,nm\,min^{-1}$, which is comparatively fast. However, the film includes carbon, which leads not only to high carbon content, but also to relatively high stress levels. If the deposited films are not annealed at high temperatures, then TEOS films thicker than about 3 µm tend to crack. Unannealed TEOS oxide etches relatively fast in both BHF and KOH. In contrast, after annealing the etch rates approach those found for thermal oxide.

Silicon dioxide can also be deposited by PECVD. The deposition is usually very fast but does produce nonconformal films. The deposition rate is usually around $200\,nm\,min^{-1}$, and the film stress can be adjusted by manipulating the deposition parameters. This means that very thick films can be deposited, with 30-µm-thick films being common. One reaction for producing $SiO_2$ using PECVD is:

$$SiH_4 + 2\,N_2O \quad \rightarrow \quad SiO_2 + 2\,N_2 + 2H_2$$

A freshly deposited film contains a lot of hydrogen and can be reasonably porous, leading to refractive indices less than 1.46. After suitable annealing, the refractive index and the etch rates in BHF and KOH resemble those of thermal oxide. Often, doped $SiO_2$ films are used; e.g., phosphorous can be added to enhance the etch rate in HF or to change the refractive index. Phosphorous doping involves incorporation of phosphopentaoxide $P_2O_5$ into the film:

$$2\,PH_3 + 5\,N_2O \quad \rightarrow \quad P_2O_5 + 3\,H_2$$

Films with high levels of phosphorous are sometimes called phosphor silicate glasses (PSG). These types of glasses can 'flow' when exposed to temperatures near 1100 °C, which is often used to planarize surfaces. The very high phosphorous contents also result in very high etch rates in HF, such as 1 µm min$^{-1}$. Sometimes boron is also added to change the refractive index or to lower the etch rate in HF. Boron incorporation follows this reaction:

$$B_2H_6 + 3\,N_2O \quad \rightarrow \quad B_2O_3 + 3\,N_2 + 3\,H_2$$

Germanium may also be added and results in an increase in the refractive index. Index-modifying compounds are often used for optical waveguide purposes where it is crucial that regions next to each other have different indices.

Films containing a mixture of silicon dioxide and silicon nitride are called silicon oxynitride SiON. These films can contain different concentrations of silicon nitride, which lead to differences in refractive index. The films have interesting stress properties because they exhibit a stress level that is the combination of the $SiO_2$ film's compressive stress and the SiN film's tensile stress. This means that stress-free films can be achieved even after annealing.

**Zinc oxide** Zinc oxide (ZnO) is a semiconductor-like material; however, because the bandgap is 3.35 eV it is often viewed as an insulator. In the crystalline form, zinc oxide shows piezoelectric behavior. When the crystal is placed in an electric field it contracts or expands, depending on the direction of the field. The opposite is also true; a potential difference develops between the two sides of the crystal if it is squeezed.

Zinc oxide can be deposited by sputtering a zinc target in an oxygen plasma. If the temperature is kept fairly high, at about 500 °C, and the substrate can be oxidized, then a crystalline film is deposited with its piezoelectrically active direction perpendicular to the wafer surface. The substrate must be oxidizable because then the ZnO molecules arrive and orient themselves correctly to the surface (zinc pointing downwards).

Both silicon and aluminum have been used as substrates for ZnO deposition. For the pure piezoelectric behavior, the ceramic lead zirconate titanate (PZT) is much more efficient; however, its use requires cooling in the presence of an electric field to orient the PZT particles correctly. Zinc oxide etches fast in most acidic solutions and should generally be shielded from aggressive chemicals.

**Borosilicate**   Borosilicate glasses are types of glasses containing sodium, potassium, boron, and calcium, among other compounds. Often, the exact composition is described by the supplier. For example, Corning 7740 Pyrex® is one type of borosilicate glass, with a specified content guaranteed by Corning, and Borofloat® from Schott is another type. Borosilicate glasses are widely used in the form of wafers for areas where transparency, high electric resistance, or special sealing properties are needed.

Some types of borosilicate glasses can be deposited as thin films by e-beam deposition. The exact composition of the wafer depends on the chamber impurities, the level of vacuum, and energy fed into the target. The films normally have a high level of sodium and can be used for thin-film anodic bonding.

**Tantalum oxide**   Tantalum oxide, TaO, one of the most chemically resistant thin-film materials ever discovered, is becoming popular. It is commonly applied by sputtering a tantalum target with an oxygen plasma, but can also be applied by LPCVD, although this process is still in its infancy due to difficulties in handling organic tantalum compounds. Tantalum oxide cannot be patterned by any wet-etching process, but can be patterned by sputtering, for instance, in a reactive ion etching chamber.

**Metal layer fundamentals**   Metal is used for many purposes in microfabrication. An overview of a few of the most common metals, their deposition methods, and uses is shown in Tab. 6.3. Metal layers in general do not adhere very well to dielectrics. However, it is often important to place a metal on dielectrics, so this adhesion issue must be overcome, often by using an adhesion layer. Such a layer must be able to stick well to both the dielectric and the metal. This approach can result in a two-layer sandwich with a thin adhesion layer and a thicker primary metal. However, the same approach can also cause problems with some metals, because the adhesion layer might allow a metal to diffuse through the layer, which would afford the primary metal access to areas where it is not desired. This problem can be avoided by inserting a third layer between the adhesion layer and the primary metal. The third layer is called a diffusion barrier. The final result is a three-layer sandwich, which is often used in metallization schemes. A fourth layer is sometimes introduced when the metal is to be covered by a dielectric. Then it is often necessary to apply another adhesion layer on top of the primary metal.

Aluminum (Al) is one of the most often used metals because it is easy to machine; it has been the staple metal used in microelectronics for more than 40 years and machinery for aluminum is readily available. Also of importance is that aluminum is fairly inexpensive even at high purity. Aluminum can be applied by sputtering, e-beam, resistive, and CVD methods. The eutectic point of the silicon/aluminum system is at 577 °C, which means that aluminum on silicon forms an alloy when heated to this temperature. This effect is known as thermomigration

**Tab. 6.3**  Various metals often used for microfabrication.

| Metal | Deposition method | Use |
|---|---|---|
| Titanium (Ti) | Evaporation<br>Sputtering | Adhesion layer, silicide former |
| Chromium (Cr) | Evaporation<br>Sputtering | Adhesion layer |
| Aluminum (Al) | Evaporation<br>Sputtering<br>CVD | Adhesion layer, conductor |
| Gold (Au) | Evaporation<br>Sputtering<br>Electroplate | Conductor, bond layer for silicon–gold eutectic bonding |
| Platinum (Pt) | Evaporation<br>Sputtering | Conductor, diffusion barrier for gold, often used as electrode in biochemical microsystems |
| Palladium (Pd) | Evaporation<br>Sputtering | Diffusion barrier for silver, can adsorb large amounts of hydrogen, silicide former |
| Silver (Ag) | Evaporation<br>Sputtering<br>Electroplate | Conductor, often used as electrode, needed for Ag/AgCl electrode systems |
| Copper (Cu) | Evaporation<br>Sputtering<br>Electroplate | Conductor, diffuses fast in silicon |
| Nickel (Ni) | Evaporation<br>Sputtering<br>Electroplate | Conductor, structural material often used for electroplated metal microstructures |

or spiking and ruins the interface. This happens to a certain extent even at lower temperatures, so the temperature should be kept below 450 °C if the metal properties of the aluminum are important.

Gold is also a very popular metal. It does not oxidize, is a good thermal and electrical conductor, is solderable, and has fairly low stress when deposited, allowing relatively thick layers to be made (up to ∼ 30 micrometers). In addition, it is fairly easy to deposit. All in all, these properties make gold the metal of choice for many applications. Unfortunately, gold does not adhere very well to most surfaces. Using gold therefore requires an adhesion layer, which is often titanium or chromium. If gold is to be used on silicon, it is necessary to use a diffusion barrier, because gold diffuses very quickly in silicon, and the eutectic point of the gold/silicon system is at the very low temperature of 363 °C. Titanium is a poor barrier for gold, and Ti/Au metallization schemes must not be subjected to very high temperatures. Chromium is slightly more stable than titanium, and Cr/Au is often used for devices in which there is no need for contacts to silicon. The diffusion barrier most often used in higher-temperature metallization schemes is platinum, which works very well even for long sintering times. However, palladium cannot be used as diffusion barrier, because it forms 'white gold', in which gold can diffuse freely into the adhesion layer even at moderate temperatures of about 300 °C.

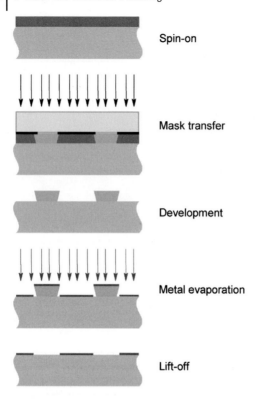

Spin-on

Mask transfer

Development

Metal evaporation

Lift-off

**Fig. 6.12** The stages of lift-off. Negative tone photoresist is spun on and patterned, giving structures with an inward slope. Then metal is evaporated to cover the wafer surface. Finally, the photoresist pattern is stripped, which also removes any metal coating on its surface.

### 6.3.4
### Lift-off

An often-used method for making metal patterns is the lift-off process (Fig. 6.12). A layer of photoresist is applied to the wafer and is structured with the appropriate pattern. The desired metal or metal sandwich is then deposited. Finally, the photoresist is stripped, and the metal that is applied to the photoresist-covered areas is removed together with the photoresist. To achieve high yield, the photoresist should slope inward. Because of the desired slope, negative tone or image reversal photoresists are usually preferred over positive tone photoresists. The metal deposition should be directed perpendicular to the wafer, making e-beam depositions popular. It is also advantageous to use a metal layer that is thinner than the photoresist coating. During stripping of the photoresist, the wafer is often subjected to ultrasonication to facilitate removal of metal flakes and to obtain more efficient stirring. Lift-off produces nice, well-defined, almost vertical sidewalls, which cannot be obtained by etching the metal. Lift-off is the preferred metallization technique for hard-to-remove metals such as platinum and tungsten.

## 6.3.5
## Silicides

Silicides are the common name for several silicon/metal alloys that share the same features. They are highly conducting, rather chemically stable, and stable to high temperatures. Many metals can form silicides, but the most popular are titanium silicide and nickel silicide. They can be sputter-deposited as alloys or deposited by chemical vapor deposition. A third option is to deposit a metal layer onto silicon (crystalline or noncrystalline) and then heat the sandwich. The silicide forms at the interface between the two layers. Normally unreacted metal layer is present after this type of treatment. For best performance, this layer must be removed. Great care should be taken when heating metal films for formation of silicide layers, because they are easily contaminated; oxygen especially causes problems, forming nonreactive islands of material. One of the great advantages of silicides is that they are generally compatible with low-pressure chemical vapor deposition (LPCVD) processes and thus may be conformally covered with an insulating dielectric.

## 6.4
## Etching Removal

Etching removal deals with approaches to removing material. Various wet (aqueous) etching approaches are described in this section. Wet etching often uses hydrofluoric acid (HF) and its mixtures with nitric acid but also potassium hydroxide (KOH) and other anisotropic wet etches (TMAH, EDP). Several other etchants often used for etching various materials are also described.

The cross-section of etches can vary (Fig. 6.13). Isotropic etchants etch with the same rate in all directions. These etchants yield profiles with cylindrical sides. Examples of isotropic etchants are HF etching of silicon dioxide and etching of sili-

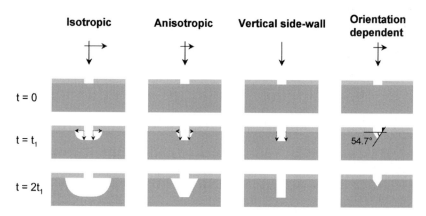

**Fig. 6.13** Etching profiles for various etches. See text for details.

con in mixtures of nitric acid and HF. For some etchants the etch rate is different in different directions, i.e. anisotropic. These etches give tapered profiles and are usually dry etches (see next chapter). Some special anisotropic etches are the vertical side-wall etches and the orientation-dependent etch. These profiles can be created by wet etching of silicon in KOH. For vertical side-walls the silicon surface must be a (110) plane and for an orientation-dependent etch giving an angle of 54.7°, the silicon plane must be (100). More details on the wet etching of silicon in KOH is given in section 6.4.4.

### 6.4.1
### Wet-etching Fundamentals

Wet etching of silicon or films on silicon involves several separate steps. First, the etchant must be transported to the surface, then the etching reactions take place, and finally the products must be transported away from the surface. Wet etching takes place on the surface of a solid in contact with a liquid solution, and in general a thin immobile boundary layer is formed right on the surface. Transport of reactants and products through this layer occur by diffusion. It is important to keep the liquid etchant as homogeneous as possible to make the etching as uniform as possible. Some etching parameters result in an etch limited by the transport of etchant to the surface or away from the surface. Such etching solutions require stirring to keep the etchant homogeneous and obtain uniform etching. Other etching parameters result in an etching process in which the overall etch rate is limited by the etching reaction; for such etches the etchant is automatically nearly homogeneous and no stirring is necessary.

### 6.4.2
### Etching with HF

HF is one of the most often used wet etchants for fabricating lab-on-a-chip systems. HF is important because it etches silicon dioxide but hardly attacks silicon itself. This means that silicon dioxide on top of a silicon wafer can be patterned by photolithography and then the pattern can be transferred by using HF (because most photoresists are especially resistant towards HF). The overall etching reaction for pure HF is:

$$SiO_2 + 6\,HF \quad \rightarrow \quad H_2SiF_6 + 2\,H_2O$$

The complex $HF_2^-$ is also present in the solution. The complex etches silicon dioxide 4.5 times faster than HF. Etching with $HF_2^-$ follows this reaction:

$$SiO_2 + 3\,HF_2^- + H^+ \quad \rightarrow \quad SiF_6^{2-} + 2\,H_2O$$

HF is used in a wide range of concentrations from about 1% to about 50%. The higher the HF concentration the higher the etch rate in silicon dioxide. The etch rate increases linearly with the concentrations of both HF and $HF_2^-$ up to 10 M. At even higher concentrations, higher-order complexes such as $H_2F_3^-$ are formed, these etch silicon dioxide even faster than $HF_2^-$. The presence of these higher-order complexes means that the etch rate increases non-linearly.

HF is a dangerous chemical because it is both highly corrosive and poisonous. It is readily absorbed through the skin, and the gas can be absorbed through the lungs. It harms skin, bones, and nerves.

Because the etch rate in silicon dioxide depends on the concentration, the etch rate changes over time in a HF bath. This is undesirable if close control of dimensions is needed or indeed if the etching time is to be somewhat constant for the same thickness of silicon dioxide on different wafers. To overcome this, a buffer system consisting of ammonium fluoride ($NH_4F$) together with HF is often used to prevent large changes in HF concentration and maintain a constant pH. Thus, the concentrations of HF and $HF_2^-$ remain constant, which in turn stabilizes the etch so that the etch rate is constant over a long time (several months). Silicon dioxide etching in the presence of $NH_4F$ follows this reaction:

$$SiO_2 + 4\,HF + 2\,NH_4F \quad \rightarrow \quad (NH_4)_2SiF_6 + 2\,H_2O$$

$NH_4F$ etchant is often termed BOE: buffered oxide etch or BHF: buffered HF; however, neither of these abbreviations is meant to specify an exact concentration. If water is allowed to evaporate from the buffer solution solid particles can appear and a film can even form on the bath surface. Adding more water may dissolve both film and particles.

### 6.4.3
### Isotropic Silicon Etch

Isotropic etching of silicon is typically performed using an etchant consisting of nitric acid ($HNO_3$) and HF. A mixture of these two chemicals etches according to this reaction:

$$6\,HF + HNO_3 + Si \quad \rightarrow \quad H_2SiF_6 + HNO_2 + H_2O + H_2$$

Silicon dioxide is etched because of the presence of HF. Silicon nitride might also be noticeably affected by this etchant. The etching solution turns yellow during etching, due to dissolved NO. Water can be used to dilute the mixture, but this leads to dissociation of nitric acid. Therefore, acetic acid is often used, in which case the etch mixture is sometimes referred to as HNA. Specific etchant compositions are sometimes referred to as CP (chemical polish) followed by a number that refers to a specific etchant composition; thus, these CP numbers are exact descriptions of the etchant. CP etches can include other additives such as iodine, bromine, or $CrO_3^-$, which helps reduce surface roughness. The etch rate can be very high, for instance, CP-4A (49.2% HF, 69.5% $HNO_3$, $CH_3COOH$ at a volume ratio of 3:5:3) etches at 35 µm min$^{-1}$; even faster etches can be obtained, although their use for micromachining is limited, due to extremely nonuniform etching and to the danger of handling such highly aggressive media.

### 6.4.4
### Orientation-dependent Silicon Etching

Some etching solutions etch silicon at etch rates that depend on the crystallographic direction. These special types of anisotropic etches are referred to as orientation-dependent etches (Fig. 6.14). The most often used etchant is KOH, which is often mixed with isopropyl alcohol (IPA). An etching solution of 28 wt% KOH in water saturated with IPA at 70 °C etches silicon in the (100), (110), and (111) planes at 0.66, 0.19, and 0.01 μm min$^{-1}$, respectively, yielding etch ratios of 66:19:1. Other compositions of etching solution and temperature can result in different etch ratios. In fact, etching without IPA increases the selectivity between the (100) and (111) planes; however, the (111) planes always etches the slowest.

Several models have been proposed to explain the large differences in etch rates, but so far no single model can explain all the experimental results. Because of this there is no generally accepted etching reaction; however, one proposed equation is:

$$Si + 2\,OH^- + 2\,H_2O \quad \rightarrow \quad SiO_2(OH)_2^{2-} + 2H_2$$

This mechanism accounts for several important experimental findings: the generation of hydrogen gas and the fact that the source of the hydroxide does not

**Fig. 6.14** Orientation dependent etching of (100) silicon in KOH. The micrograph shows the inverted pyramid structure typical of ODE.

**Fig. 6.15** Convex corner underetch using (100) silicon, etched in KOH saturated with isopropyl alcohol. The masking layer is 120 nm of stoichiometric LPCVD silicon nitride.

matter. Experiments have been carried out with for instance LiOH, NaOH, CsOH, NH$_4$OH, and KOH, and all result in an orientation-dependent etch.

The most important point regarding orientation-dependent etches is the large etch rate differences, which are often used to form channels and pits in silicon. In (100) silicon wafers, the (111) planes form right angles with each other. This leads to structures etched in (100) silicon being bounded by (111) planes (for sufficiently long etches), which produces rectangular structures.

Corners are an important issue when dealing with KOH-based etchants, because convex corners are heavily attacked (see Fig. 6.15). The shape of convex corners during etching is determined by the fastest-etching planes, and the shape of concave corners is determined by the slowest-etching planes.

## 6.4.5
### Common Orientation-dependent Etchants

Several orientation-dependent etchants exist; the best studied is KOH. The major shortcomings of KOH are the fairly rapid etching of silicon dioxide, large amount of gaseous hydrogen evolved, and the leftover alkali potassium ions, which can significantly reduce the quality of dielectric layers. This quality reduction is mainly of importance in electronic components and not so much in applications where mainly the structural or resistive properties of dielectrics are important. An etchant composed of ethylenediaminepyrocatechol (EDP) and pyrazine is an often-used alternative to KOH, which etches silicon dioxide at one tenth the rate of KOH and generates fewer bubbles. Unfortunately, it is toxic and decomposes quickly, especially in the presence of oxygen. Nevertheless it has until recently been the second most popular orientation-dependent etchant, a position that is now rivaled by tetramethylammonium hydroxide (TMAH). TMAH is gaining popularity because it is fairly easy to handle, fully compatible with electronics fabrication, and also because if it is suitably doped with other materials (such as boric acid and silicon). TMAH does not attack aluminum. Furthermore, the masking properties of silicon dioxide and silicon nitride are even better for TMAH than for KOH or EDP. Actually, the masking properties of $SiO_2$ become so good that even a natural oxide on silicon only a few nanometers thick requires a considerable amount of time to etch. Other kinds of orientation-dependent etches exist, including hydrazine–water, ammonium hydroxide–water (AHW), tetraethylammonium hydroxide (TEA), and other hydroxides (LiOH, NaOH, CsOH). Each of these has several interesting properties, but mostly they are avoided because they do not offer enough advantages or are too expensive or dangerous to work with; for instance, hydrazine–water is not only corrosive but also is a suspected carcinogen and explosive at high concentrations.

## 6.4.6
## Other Etchants

Several wet etchants are summarized in Tab. 6.4. All are isotropic and result in cylindrical edge profiles.

**Tab. 6.4** Various wet etchants

| Target material | Etchant composition and temperature | Etch rate |
|---|---|---|
| $Si_3N_4$ | $H_3PO_4$ (85 vol%) $H_2O$ (15 vol%) 160 °C | 0,3 nm min$^{-1}$ |
| Au | Tri-iodide: 4 g KI 1 g $I_2$ 40 mL $H_2O$ | 1 μm min$^{-1}$ |
| Pt | 1 mL $HNO_3$ 7 mL HCl 8 mL $H_2O$ | 50 nm min$^{-1}$ |
| Al | 1 mL $HNO_3$ 4 mL $CH_3COOH$ 4 mL $H_3PO_4$ 1 mL $H_2O$ | 35 nm min$^{-1}$ |
| Al (2% Si) | 1 mL $HNO_3$ 1 mL $CH_3COOH$ 16 mL $H_3PO_4$ 2 mL $H_2O$ 50 °C | 650 nm min$^{-1}$ |
| Ti | 1 mL $H_2O_2$ 1 mL HF 20 mL $H_2O$ | 900 nm min$^{-1}$ |

6.4.7
**Effects of Not Stirring a Transport-limited Etch**

If one neglects stirring when using an etching reaction that is reaction-limited, several effects may occur. Because the reactants are consumed, the concentration of etchant becomes lower near the surface compared to the rest of the liquid. Some etchings occur more slowly, for instance, HF etching silicon dioxide; resulting in feature-size-dependent etching. Other etchants might etch faster or even change the profile of the microstructure, for instance, KOH etching silicon, where for some concentrations of KOH a lower concentration results in faster etching and less selectivity between the (100) and (111) directions. Also for KOH, lack of stirring might lead to formation of hydrogen bubbles, which effectively shield the etch and may form starting points for the formation of hillocks and other rough spots.

6.5
**Dry Etching**

Several etching processes that do not use liquids to supply the etchant are commonly referred to as dry etching techniques. The most important ones, such as reactive ion etching and plasma ashing, use gases in a plasma state (see Fig. 6.16).

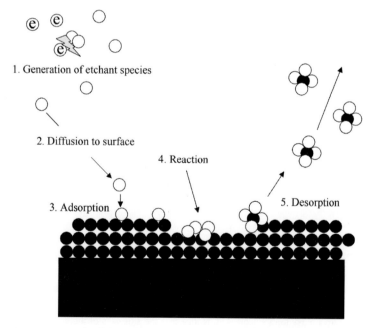

**Fig. 6.16** Principle of dry etching.

This section describes plasma based etching techniques in some detail and mentions other dry etching techniques as well. Dry etching techniques have several benefits. Waste from the processes is often easier to dispose of, and the machinery is much safer to use than that used with most wet etches.

### 6.5.1
### Plasma Etching Fundamentals

Plasma etching uses low pressures (10–500 mTorr, 1 mTorr=0.1333 Pa), low temperatures (cryogenic to $\sim150\,^\circ$C), and radio frequency power supplies (13.56 MHz). Under these conditions, a fraction of gas molecules dissociate into electrons and ionized radicals. The charged species may be affected by an electric or magnetic field and may be accelerated. The electrical power fed into the system varies from a few watts to kilowatts, depending on the desired etch properties. Much of the theory behind plasma etching is poorly understood and much effort is spend on optimizing recipes for different purposes. Several fundamental concepts regarding the etch and characterization of the etch are presented in this section.

*Chemical etching* occurs in a plasma when some of the radicals in the gas react with the wafer, forming volatile compounds. The same considerations regarding reaction kinetics and mass transport as for the wet etching processes apply also to this type of etching. First, the radical must reach the surface and adsorb, then react, and finally desorb and be transported away (Fig. 6.16). For most etching set-ups the chemical etch rate is higher when the electrical power is higher (leading to higher dissociation efficiency), the pressure is higher, or the gas flows are high-

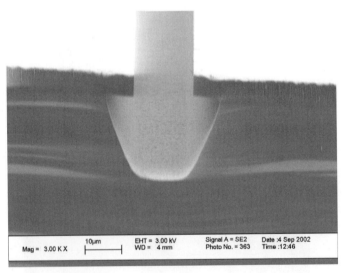

| Mag = 3.00 K X | 10µm | EHT = 3.00 kV | Signal A = SE2 | Date :4 Sep 2002 |
| | | WD = 4 mm | Photo No. = 363 | Time :12:46 |

**Fig. 6.17** Example of mask undercut and sloping sidewalls. Silicon dioxide is used as a mask for a silicon etch. Notice that the sidewalls of the silicon dioxide are close to vertical.

er (a greater supply of reactants). Chemical etching is usually isotropic and requires special passivation to avoid cylindrical wall shapes.

In *physical etching* atoms bombard a surface carrying such high momentum that, for each incoming atom, more than one target atom is knocked out of its position in the solid. Physical etching is also known as sputtering. The etch rate of physical etching is higher when the electrical power is higher (due to higher energy of the individual atoms) or the pressure is lower (longer mean free path length). Physical etches typically produce anisotropic profiles, because most atoms strike the surface at angles close to 90° (Fig. 6.17).

In physicochemical etching, neither physical nor chemical etching predominates, and energy supplied by the physical etch enhances the reaction rate of the chemical etch. Physicochemical etch is often called reactive ion etching (RIE) and is used for several of the most useful plasma etching techniques. However, unless the etch products are volatile they may redeposit or require high-energy atoms to sputter off. This is one reason for the poor performance of plasma etching of borosilicate glasses.

*Sidewall passivation* occurs when certain types of polymers (mainly fluorocarbons and fluorosilanes) present on sidewalls suppress the chemical etching. These Teflon®-like polymers can be made in the plasma during processing if both carbon and fluorine are present. Generally, such a polymer covers the surface in a conformal manner. In areas where the fluoropolymer is exposed to the physical etch, it is often removed due to sputtering. In this fashion, in some regions both physical and chemical etching occur, while other regions are shielded from the physical etch and have fluoropolymer to passivate the chemical etch. The latter regions are not etched at all, resulting in the fabrication of vertical sidewalls.

*Masking* is required to transfer the patterns from a mask to the substrate to be etched. The plasma etches the masking layer as well, which often means that the mask erodes during etching. This in turn means that the selectivity between the masking layer and the layer to be etched must be taken into account when deciding on the mask layer thickness. Metals, thin-film dielectrics, silicon, and photoresist can all be used as masking layers. The masking scheme is usually chosen on the basis of simplicity in the fabrication and often, but to a lesser extent, on whether the masking layer is slowly etched or not. Factors to take into account when evaluating masking simplicity are the application and removal of the mask layer, ease of patterning the mask layer, and etching influence of the mask layer. Because of charge effects, some dielectrics and also some photoresists can yield a different etch profile than a mask based on metal.

*Loading* describes changes in the etch uniformity due to differences in etched area. Macroloading deals with differences between two wafers having different masks. One wafer can, for instance, have an area of 10% that should be etched and another wafer may have an area of 25% to be etched. If the same machine and processing recipe is used but the etch rate is different for these two wafers, this is called a macroloading effect. If the recipe, however, can achieve the same etch rates regardless of the area to be etched, then the procedure does not exhibit macroloading.

Loading can also happen on a more local scale, for instance, if two 50-μm-diameter holes should be etched, and one of the holes has no other structures within 1 cm whereas the other hole has large structures nearby which will also be etched. If these two holes do not etch to the same depth, it is said to be due to microloading. This means that local changes in the etchant composition result in changes in the etch rate. Process optimization can often reduce microloading. However, the optimized process is usually very specific for a given mask layout and is probably even more subject to macroloading effects.

*Lag* occurs when features of different sizes do not etch at the same rate. Lag is positive when smaller structures etch faster than larger structures and negative when the opposite occurs. Lag is an extreme example of microloading.

*Aspect ratio* is the etch depth divided by the width of the feature and is interesting only for anisotropic etches. A high aspect ratio implies deep narrow trenches or holes. It is important to know what kinds of structures were used when measuring the aspect ratio, because of loading effects.

### 6.5.2
### Plasma Etching Setups

In a *plasma asher*, molecular oxygen ($O_2$) dissociates to atomic oxygen (O), which is much more reactive. The atomic oxygen then attacks organic films, such as photoresists. Etching in a plasma asher is isotropic. Plasma ashing is popular because high throughput can be obtained. The most often used plasma asher is the barrel asher, in which the wafers are placed in a boat surrounded by a Faraday cage. A low-pressure oxygen plasma is created and, because of the Faraday cage, the oxygen mainly diffuses in the reactor; therefore, only chemical etching takes place, and no appreciable amount of physical etching occurs. Because of the plasma, many high-energy photons are created, which may lead to device damage on the wafers.

*Reactive ion etching* (RIE) is often performed in a reaction chamber with parallel plate electrodes. The wafer to be etched is placed on the cathode, and RF power is applied between the upper and lower electrodes. A couple of capacitors are used to match the impedance of power supply and plasma. The etch gases present dissociate to a certain extent (depending on the parameters), and the fast electrons and slower positive ions are accelerated by the AC field (Fig. 6.18). The electrons are much faster than the positive ions, so a large fraction of them are lost to the walls and upper electrode. This process results in a net positive plasma potential. The time average (rms) of this potential is called the bias.

In general, the higher the bias voltage, the higher the energy of the sputtering atoms and the larger the mask erosion. Reactive ion etching allows for etching many different materials and, depending on the parameters, allows a large variation in etch profiles. RIE etches are comparatively complex and hard to predict, because of the number of parameters involved, which all influence the etch rate and loading. What also makes RIE etches challenging is that the history of the chamber is important. Something that was etched a couple of days ago might in-

**Fig. 6.18** Mechanism of reactive
ion etching of silicon with sulfur-
hexafluoride.

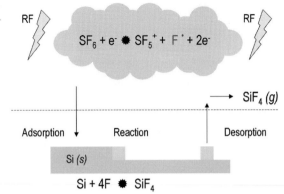

**Fig. 6.18** Mechanism of reactive
ion etching of silicon with sulfur-
hexafluoride.

fluence the etch performed today. Thus processes are usually optimized for specific masks on specific machines and a conditioning step is inserted to make certain that the chamber properties are the same from etch to etch.

Processes designed specifically for deep etches into silicon are called deep reactive ion etching (DRIE). The two main approaches are the cryogenic process and the Bosch process (Fig. 6.19). They differ considerably from each other in some parts of the setup, but both use an inductively coupled plasma (ICP). ICP is used to create a magnetic envelope inside the etch chamber, which reduces the loss of charged species to the surroundings and helps to achieve and maintain a higher plasma density. ICP systems generally have better uniformity, better selectivity to mask layers, higher etch rates, and they provide the opportunity to achieve deeper etches, as compared to RIE. On all accounts, inductively coupled plasma deep reactive ion etchers perform better than traditional parallel plate RIE when etching silicon.

Cryogenic DRIE uses a setup much like that of a standard RIE with parallel plates; however, the stage for the wafer is cooled much more efficiently and to much lower temperatures. Cooling the substrate lowers the surface mobility of reactive ions, which means that ions hitting the surface do not migrate. This leads to very little etching of the sides of a trench at cryogenic temperatures, which should be as low as possible to achieve the highest anisotropy. Cryogenic tempera-

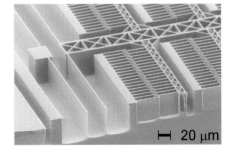

**Fig. 6.19** An example of deep reactive ion
etching (DRIE). Notice the vertical sidewalls
and the high depth-to-width ratio of the
structures. This structure was etched by
the Bosch process. (Picture courtesy:
S. Jensen).

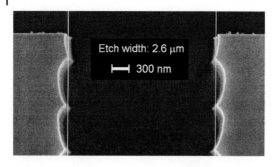

Etch width: 2.6 µm

⊢⊣ 300 nm

**Fig. 6.20** SEM micrograph showing a close-up view of scallops generated during DRIE using the Bosch process. (Picture courtesy: M. Balslev).

tures are often obtained by several stages of cooling. The cooling system must be able to dissipate the heat generated by the etching process.

Bosch DRIE uses cycles of etching and sidewall passivation. The cycle time is usually 5–30 s. The cycles result in uneven etching of the sidewall, creating scallops, the size of which depend on the cycle time. Typical scallop depths are a few hundred nanometers (Fig. 6.20). The Bosch process is named after the company where the inventors of the process worked. This method can give etch rates of up to 10 µm min$^{-1}$ in silicon using photoresist as mask layer.

### 6.5.3
### Etch Gases

Many different etch gases have been used over the years. For etching of photoresist, oxygen is used. If other materials are to be etched, it is usually necessary to employ gases that contain halogens such as bromine, chlorine, or fluorine. A free excited halogen in plasma is highly reactive and can be used to etch most materials. Chlorine and fluorine attack almost all materials. Aluminum is an important exception because fluorine does not etch aluminum. Here, only certain commonly used gases based on fluorine gases are described.

Several of these compounds are themselves CFC (Freon) compounds, or CFC compounds might be waste products of the reactions. Care must be taken when disposing the waste (often in gas form), because CFC compounds in the earth's atmosphere attack the ozone layer. Although various gases are used, some of the basics remain the same. Fluorine from an etch gas is used for etching most solids and for fluorocarbon sidewall passivation. A typical example of fluorine etching is the etching of silicon nitride:

$$Si_3N_4 + 12\,F \quad \rightarrow \quad 3\,SiF_4 + 2\,N_2$$

Only the free fluorine is active in this etch. Carbon-supplying gases are used for sidewall passivation purposes or to form CO or $CO_2$ from etching oxides, as when etching $SiO_2$ in a $CHF_3$ plasma:

$$3\,SiO_2 + 4\,CHF_3 \quad \rightarrow \quad 2\,CO + 2\,CO_2 + 3\,SiF_4 + 2\,H_2$$

**Tab. 6.5** Common gases used for reactive ion etching

| Name | Formula | Use |
|---|---|---|
| Oxygen | $O_2$ | Oxygen radicals etch photoresist |
| | | Can displace F in CF and SF compounds |
| Sulfur hexafluoride | $SF_6$ | F source |
| | | Sulfur atoms are heavy and can sputter effectively |
| Tetrafluoromethane | $CF_4$ | F source |
| (Freon 14) | | C source for sidewall passivation |
| Trifluoromethane | $CHF_3$ | F source |
| (Freon 23) | | C source for sidewall passivation |
| Nitrogen | $N_2$ | Diluent |
| | | Stabilizes oxygen plasmas |
| Helium | He | Diluent |
| | | Plasma stabilizer |
| Argon | Ar | Diluent |
| | | Ar atoms are heavy and can sputter effectively |

Carbon monoxide and carbon dioxide are both very stable and are thus carried away by the gas stream without dissociating. Oxygen can be added to etch organics or to replace fluorine in CF and SF compounds, leading to more available fluorine. Some gases can be used to dilute the etch gas without themselves taking part in the etching. Often, however, a seemingly inert gas changes several parameters of the plasma chemistry. Nitrogen, for instance, can be used to dilute gases. Nitrogen itself does not generally participate in etching any material, but it can facilitate energy transfer from the electrical field to oxygen. Thus, oxygen plasmas tend to have a larger fraction of dissociated oxygen when nitrogen is present. Some etch gases include heavy atoms (such as sulfur or argon), which sputter the target.

### 6.5.4
### Laser-assisted Etching

Chlorine etches silicon; however, the etch rate is appreciable only if the chlorine is ionized (as in RIE) or the temperature is high. One method to achieve very high temperatures is to use an argon-ion laser at 488 nm. The radiation is readily absorbed by silicon and locally heats the material. The argon-ion laser beam has a gaussian intensity profile and, by using optical focusing, the resolution of this laser-assisted etching can be $\sim 500$ nm. Laser-assisted etching can be used for creating 3D structures. It can also be used for prototyping purposes to make patterns in a masking layer that may then be extruded by another etch method.

**6.6**

**Heat Treatment**

6.6.1

**Thermal Oxidation**

One of the most important reasons for the widespread use of silicon is the ability to make silicon dioxide ($SiO_2$) on the surface. Silicon dioxide is a good electrical insulator as well as being mechanically and chemically stable (Tab. 6.6); it is a major constituent in all kinds of glasses, from mundane window glass to high quality quartz. In fact, quartz is the crystalline form of silicon dioxide found in nature. Thermal oxide is silicon dioxide created by oxidation of part of the silicon wafer into silicon dioxide. Silicon is oxidized at elevated temperatures (550–1200 °C). Either oxygen ($O_2$) or water vapor ($H_2O$) can be used as the oxygen source (Fig. 6.21). Oxidation by $O_2$ is often called 'dry' and oxidation by water vapor is conversely called 'wet'.

Oxidation is usually carried out at atmospheric pressure in a furnace consisting of a quartz tube surrounded by heating elements. The silicon wafers are placed in the tube in a boat, normally standing upright. At one end of the tube, gases are fed; the flow is carefully controlled, assuring laminar flow inside the tube. Turbulence in this type of furnace would cause uneven gas coverage of the wafers, leading to uneven layers; it could also stir up any dust that might be present in the furnace. Thus, laminar flow is important in obtaining uniform layers without contamination by particles.

The silicon dioxide is grown from the surface of the wafer, and the growth rate of the initial thin layer is limited by the surface reaction. As the silicon dioxide becomes thicker, the oxygen needed for oxidation must diffuse through the silicon dioxide layer, and the growth rate is thus limited by diffusion. It is important to note that the silicon dioxide is created at the interface between silicon and $SiO_2$, meaning that the silicon dioxide that formed first is on the surface of the wafer.

Oxidations using water vapor result in thicker layers than when oxygen alone is used. The growth temperature is important; and the higher the temperature, the higher the growth rate (Fig. 6.22). Of secondary importance is the crystallographic

**Tab. 6.6** Basic properties of silicon dioxide.

| Property | Value |
| --- | --- |
| Density (g cm$^{-3}$) | 2.2 |
| Refractive index | 1.46 |
| Dielectric strength (V nm$^{-1}$) | >1 |
| Energy gap (eV) | 9 |
| Thermal conductivity (W cm$^{-1}$ K$^{-1}$) | 0.014 |
| Thermal expansion coefficient (K$^{-1}$) | $5 \times 10^{-7}$ |
| Resistivity (300 K, $\Omega$cm) | $10^{14}$–$10^{16}$ |
| Melting point (°C) | 1705 |

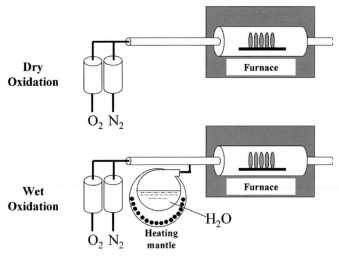

**Fig. 6.21** Thermal oxidation furnaces.

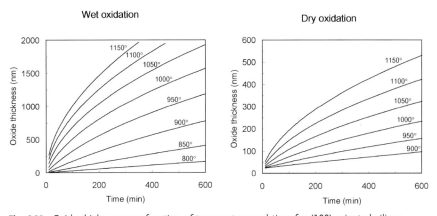

**Fig. 6.22** Oxide thickness as a function of temperature and time for (100) oriented silicon.

direction of the silicon substrate – (111) material oxidizes faster than (100). The properties of thermal silicon dioxide layers are relatively similar except for the interface properties. When high quality electrical interfaces are needed, oxidations should be done with pure oxygen and at temperatures that are not too high. The thickness of the film, $x_{ox}$, as a function of the oxidation time $t$ is frequently modeled by a linear–parabolic empirical expression developed by Deal and Grove:

$$x_{ox}^2 + Ax_{ox} = B(t + \tau) \tag{6.4}$$

$$\tau = \frac{1}{B}x_i^2 + \frac{A}{B}x_i \tag{6.5}$$

**Tab. 6.7** The linear and parabolic rate constants as a function of absolute temperature ($k$ is the Boltzmann constant).

| Oxidation | Orientation | $B/A$ ($\mu m \ h^{-1}$) | $B$ ($\mu m^2 \ h^{-1}$) |
|-----------|-------------|--------------------------|--------------------------|
| Dry | (100) | $3.708 \times 10^6 \exp(-2.00 \ \text{eV} \ k^{-1} T^{-1})$ | $7.720 \times 10^2 \exp(-1.23 \ \text{eV} \ k^{-1} T^{-1})$ |
|     | (111) | $6.230 \times 10^6 \exp(-2.00 \ \text{eV} \ k^{-1} T^{-1})$ | |
| Wet | (100) | $7.512 \times 10^7 \exp(-2.05 \ \text{eV} \ k^{-1} T^{-1})$ | $4.200 \times 10^2 \exp(-0.78 \ \text{eV} \ k^{-1} T^{-1})$ |
|     | (111) | $1.262 \times 10^8 \exp(-2.05 \ \text{eV} \ k^{-1} T^{-1})$ | |

where $B$ is the parabolic rate constant and $B/A$ is the linear rate constant (Tab. 6.7). The parameter $\tau$ is a time displacement accounting for any initial oxide layer of thickness $x_i$. The model can be solved to give the oxide thickness as a function of oxidation time:

$$x_{ox} = \frac{A}{2}\left(\sqrt{1 + \frac{4B(t + \tau)}{A^2}} - 1\right) \tag{6.6}$$

This model is relatively good for wet oxidation; however, it must be modified for dry oxidation. Here, an initial layer of oxide grows comparatively fast, and so the initial oxide thickness must be modified by adding 23 nm to it to obtain $\tau$.

If all silicon dioxide is removed from a silicon surface, a layer of oxide starts to grow even at room temperature, if the silicon is exposed to atmospheric air. This oxide layer is called native oxide and has a thickness between 0.6 nm and ~1.5 nm, depending on the time it has been exposed to air.

The oxidation process is influenced by various impurities in the substrate. For instance, high levels of phosphorous in the silicon lead to much thicker oxide layers than expected. In addition, the properties of a phosphorous-rich silicon dioxide film are different from the standard film. Of particular importance is the high etch rate of such a film by HF, which can be up to thirty times faster than for the standard oxide.

## 6.6.2
## Diffusion

Dopants can be introduced into the silicon in several ways, using thermal processes. A process called drive-in involves a layer containing the dopant material located on top of the wafer. The wafer and this layer are heated to facilitate diffusion, which drives dopants from the source layer into the silicon substrate. Another approach is to implant dopants by ion implantation and then activate the dopants by heat.

Dopant atoms can diffuse into and within the silicon crystal. The phenomenon of diffusion is explained in detail in section 3.2. Here, only the essential facts are repeated. The process of equalizing differing concentrations of a certain type of

particle within a gas, liquid, or solid is called diffusion. In doping, diffusion is interesting with respect to the spreading of dopant atoms in a silicon crystal. The
diffusion speed is characterized by the diffusion constant $D$, which depends on
the dopant–crystal combination and the temperature. The larger the concentration
differences of dopants within the crystal and the higher the temperature, the faster the dopants diffuse within the host crystal. The average distance moved in a
time period $t$ is called the diffusion length $L_d = \sqrt{Dt}$.

Two of the dopants most often used, boron and phosphorous, have comparatively low diffusion coefficients. For instance, at $1000\,^\circ$C the diffusion coefficients
are approximately the same: $10^{-14}\,\mathrm{cm^2\,s^{-1}}$. For diffusion for 4 h at $1000\,^\circ$C the diffusion length is thus 0.12 μm.

Two distinctly different diffusion profiles can arise from different initial and
boundary conditions. If diffusion is carried out with a constant total amount of
dopant, the resulting concentration profile has a gaussian shape:

$$C(x,t) = \frac{S}{\sqrt{\pi Dt}} \exp\left(-\frac{x^2}{4Dt}\right) \qquad (6.7)$$

where $x$ is the distance from the surface into the silicon crystal, $t$ is the time, and
$S$ is the total amount of dopant.

Alternatively, if the surface concentration is constant, the concentration profile
after diffusion follows a complementary error function expression:

$$C(x,t) = C_s \mathrm{erfc}\left(\frac{x}{2\sqrt{Dt}}\right) \qquad (6.8)$$

where $C_s$ is the concentration at the surface. The two expressions both yield
rounded, soft profiles. Achieving very deep doping requires high temperatures (up
to $1200\,^\circ$C) and very long diffusion times (several days).

Doping by predeposition involves placing a layer containing many dopant
atoms on the wafer. Thermal processing can create such a layer by thermal oxidation in an atmosphere containing the dopant atoms. The gas $BBr_3$ is commonly
added to create a boron-rich layer. For phosphorous, liquid $POCl_3$ is often used;
the oxygen for oxidation bubbles through a tank containing $POCl_3$ at room temperature. Some $POCl_3$ vapor is carried with the oxygen into the furnace and creates a phosphorous-rich film.

### 6.6.3
### Annealing

Annealing is a thermal treatment under a more-or-less inert atmosphere. The goal
is to create some physiochemical changes. Annealing often means that a system
comes closer to thermal equilibrium. A thermally grown wet silicon dioxide film,
for example, contains a large concentration of water molecules. Annealing such a
film in an inert (nitrogen or argon) atmosphere removes the water molecules.

The composition of the film does not change but its physical properties do. Another effect of annealing is to control the total stress of deposited films. Many films do not have a thermal expansion coefficient equal to that of silicon, which means that upon cooling, the mismatch in expansion coefficient results in a stress component. By choosing a specific maximum temperature, this part of the total stress can be controlled.

6.6.4
## Wafer Bonding

Wafer bonding provides a means of obtaining sealed channels, which are used in most microanalytical devices, especially for handling liquids. Wafer bonding is thus a fundamental, but also very critical, part of a typical microfabrication process, because the process often fails, resulting in leaky channels. This section provides a short introduction to one of the most often used bonding methods, *anodic bonding*. Another widely used bonding technique, called thermal fusion bonding, is described in Chapter 7, because it is mainly used for bonding two dielectric materials.

### Physical mechanism of anodic bonding

A structured silicon substrate is typically bonded to a borosilicate glass cover plate by means of field-assisted thermal bonding, also known as anodic or electrostatic bonding.

A setup for anodic bonding consists of a vacuum chamber, two electrodes, and heat plates for temperature control. The assembled wafer pair is placed between the electrodes, and an electrical field of 200–1000 V is applied across the wafers. The glass substrate is biased negatively (cathode) with respect to the silicon substrate (anode). The temperature is typically 200–450 °C. The principle behind the bonding mechanism is sketched in Fig. 6.23.

Some glass wafers have been developed specially for this process; having a high content of various ions such as sodium and boron (hence the name borosilicate), which are added to make the substrate slightly conducting. When the field is applied, the sodium ions ($Na^+$) in the borosilicate glass substrate migrate towards the cathode, and oxygen ions ($O^{2-}$) migrate towards the anode. The migration results in a negative space charge region in the glass interface towards the silicon substrate, and so the two wafers are electrostatically attracted to each other. This attraction takes place locally at the wafer interfaces and thus is more efficient for covering over defects and particles than for pressing two wafers together with an external load, as is used for thermal fusion bonding of glass and polymers (Chapters 7 and 8).

The presence of oxygen ions at the wafer interface results in oxidation of the bottom substrate and hence establishes a bond between the two wafers, so the wafers do not fall apart when the electrical field is turned off. The bond is stronger than either of the two materials, and an attempt to break the bond fractures the glass or the silicon substrate. The very strong bond that can be achieved with this

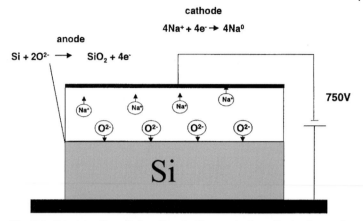

**Fig. 6.23** Principle of anodic bonding. A silicon and a borosilicate glass wafer are placed on top of each other, and a high voltage (here, 750 V) is applied across both wafers. The electric field inside the wafers causes migration of oxygen and sodium ions, which in turn results in an attractive force between the two wafers. Upon heating, very strong bonding occurs. Graphic from A. Bertold, et al. Eurosensors XIII, p. 975 (1999).

method is the reason for its extensive use. Ions do not move if the glass is made of fused silica (also called quartz), so no bonding occurs.

Anodic bonding has severe limitations, because this method is specific to the silicon and borosilicate material combination. If another material combination is desired, anodic bonding might be achieved only by additional processing of one of the substrates. Two silicon wafers can be bonded anodically if a layer of sodium-rich borosilicate glass is deposited on one of the wafers. Two glass substrates cannot be anodically bonded without using at least one intermediate layer [5]. It is furthermore impossible to anodically bond a pure silica substrate (quartz) to a silicon wafer, because of the low concentration of sodium ions in this material. It is hence impossible to take advantage of the good UV transparency (for optical detection) of quartz relative to borosilicate glass (boron in the glass network causes strong UV absorption). These shortcomings have resulted in great interest in direct bonding methods, such as thermal fusion bonding (see Chapter 7).

**Practical information**
Before bonding, the wafers are typically rinsed in a 'Piranha solution', which is a 1 to $\sim 3$ mixture of hydrogen peroxide ($H_2O_2$) and sulfuric acid ($H_2SO_4$). A so-called RCA1 solution can also be used, which consists of water, hydrogen peroxide ($H_2O_2$), and ammonium hydroxide ($NH_4OH$) in, e.g., a 400:100:70 ratio. This rinsing should never be neglected, on the grounds that a wafer is clean enough: this initial rinsing is not merely a cleaning step, but also a way of making the wafer surfaces hydrophilic, which significantly improves the quality of the bond and might make the difference between obtaining a leaky or a sealed channel network.

## 6.7
## References

BERTOLD, A. et al., Eurosensors XIII, p 975 (1999)

KOVACS, G. T. A., *Micromachined Transducers Handbook*. McGraw-Hill, 1998

MADOU, M., *Fundamentals of microfabrication*. Boca Raton, CRC Press, 1997

MAY, G. S., SZE, S. M., *Fundamentals of semiconductor fabrication*. John Wiley & Sons, 2004

SZE, S. M., *Semiconductor devices. Physics and technology*. John Wiley & Sons, 1985

# 7
# Glass Micromachining

Daria Petersen, Klaus Bo Mogensen, and Henning Klank

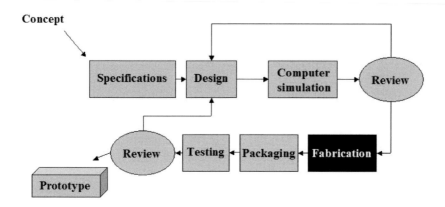

Glass can be characterized as a monolithic noncrystalline solid consisting mainly of silicon dioxide ($SiO_2$). Glass often contains additives or impurities that modify properties such as its mechanical stability and the glass transition temperature. An endless variety of glasses with different chemical compositions and physical properties are in use today. One can find optical glasses (filters and lenses), light sensitive and photochromic glasses, x-ray and gamma-ray absorbing glasses, glasses with coloring and discoloring agents, with luminescence and fluorescence effects, in applications like glass pH electrodes, glass lasers, optical fibers, and many others. The classification of glasses according to their optical properties is given in *Optical Properties of Glass* by Fanderlik [1], who gives attention to glasses exhibiting special optical properties and to the dependence of glass properties on its chemical composition, temperature, and thermal history.

The use of glass instead of silicon in μTAS applications is prompted by the unique properties of glass, e.g., glass is resistant to many chemicals, optically transparent (which allows for optical detection and visual inspection), and dielectric (and can therefore withstand the high voltages used in electrokinetically driven flows and separations). Other advantages of glass are its hardness, high thermal stability, and relative biocompatibility, which broadens its range of applications to DNA separations, enzyme reactors, immunoassays, and cell biology. For μTAS applications,

*Microsystem Engineering of Lab-on-a-chip Devices*
O. Geschke, H. Klank, P. Telleman
Copyright © 2004 Wiley-VCH Verlag GmbH & Co. KGaA, Weinheim
ISBN: 3-527-30733-8

Borofloat glass (and Pyrex®) and quartz wafers are mostly used, because they are compatible with many cleanroom processes. Quartz, the crystalline form of silicon dioxide, is an abundant mineral appearing in many forms and colors and occurring as grains (sand), in masses (chalcedony, carnelian, and jasper), and in hexagonal crystals as amethyst and rock crystal. Quartz is very hard, has a low thermal expansion coefficient, and transmits light from about 180 nm (UV range) to 4.5 μm (infrared). Being optically transparent to UV light, quartz can be employed for UV-absorption detection on microchips. However, crystalline quartz of optical quality is very expensive. Moreover, bonding of quartz to silicon or glass is difficult, because of the high melting point of quartz, 1713 °C. Quartz that has been melted and cooled to form amorphous glass, so-called fused silica, is easier to use in μTAS applications.

Borofloat glass shares many of the beneficial properties of quartz, i.e., Borofloat glass is chemically resistant, a dielectric, and optically transparent (although not down to the UV range like quartz and fused silica). Borofloat glass is significantly cheaper than quartz and can be bonded to silicon and glass rather easily. Borofloat glass is therefore extensively used in μTAS applications and is in the focus of this chapter.

Some traditional silicon microfabrication techniques, such as photolithography and wet chemical etching, have been adapted to glass processing. Nevertheless, micromachining of glass is less versatile than that of silicon, due to its noncrystalline structure and the limited experience with glass as a material for microsystems.

## 7.1
### Wet Chemical Etching

Wet chemical etching for fabricating microfluidic networks in glass involves the use of traditional silicon-processing techniques such as photolithography and wet chemical etching. The glass substrate is spin coated with a 3–6-μm-thick film of photoresist. UV exposure through a mask and development of the photoresist provide the desired pattern of microfluidic channels. Because etch rates in glass are relatively low, etching times are typically fairly long. One problem in using photoresist as an etching mask in glass etching is its poor adhesion to glass. The photoresist layer tends to lift off during etching, allowing the etchant to etch the glass substrate indiscriminately. To prevent lift-off problems with photoresist, other masking materials are often used for glass etching, e.g., chromium, gold, aluminum, and polysilicon. The desired pattern is transferred to the mask layer by photolithography, as described in section 6.2.1.

Wet etching of glass is mostly done with hydrofluoric acid (HF) or buffered hydrofluoric acid (BHF). The chemical reaction involved in wet etching glass is

$$SiO_2 + 4\,HF \quad \longrightarrow \quad SiF_4 + 2\,H_2O$$

The etch rate of Pyrex® glass in 40% HF at room temperature (22 °C) is about 3.4 μm min$^{-1}$, and the etch rate of Pyrex® glass in BHF (5% HF) at room temperature is only about 0.04 μm min$^{-1}$. Adding 5–10% HCl to BHF (5% HF) increases the etch rate of Pyrex® glass to 0.23 μm min$^{-1}$ at room temperature.

Because of the highly toxic nature of the HF used in glass etching, special safety precautions must be taken. Face shields and proper chemical-resistant gloves must be used at all times.

Glass is an amorphous material; therefore, the wet etching is always isotropic, meaning that curved geometries can be obtained. Sharp corners and high aspect ratios cannot be achieved by wet etching of glass. Inherent to isotropic etching is under-etching of the mask, resulting in channels wider than the pattern in the photoresist would predict. This should be taken into account in designing a mask. For example, if the desired channel dimensions are 100 μm wide and 25 μm deep, the channel width on the lithography mask should be only 50 μm, because the etchant removes glass equally in all directions. While etching to a 25-μm depth, the glass is also etched approximately 25 μm on each side, for a total channel width of 100 μm (50 μm + 2×25 μm = 100 μm). Under-etching tends to result in high surface roughness of the channel walls.

An exception to the rule that glass is always etched isotropically is a fabrication process that employs photostructurable glass [7]. The glass is masked by a photo-lithographic process and then chemically etched.

## 7.2
## Reactive Ion Etching (RIE) of Glass

Dry chemical etching of glass is also possible. The etch rate depends on the RIE equipment, glass composition, and the process settings. For glass etching, a $CF_4/O_2$ plasma is typically used. For pure silicon dioxide, the RIE etch rates can be at least 200 nm min$^{-1}$, but probably higher if the equipment is optimized for this purpose. This is slow compared to the fast, inexpensive batch microfabrication that can be obtained with wet chemical etching.

## 7.3
## Laser Patterning

Lasers can be used to apply large amounts of energy in small areas. With sufficient amounts of energy, material can be physically removed. Laser patterning of various glasses has been studied much recently. Laser micromachining can be used routinely to fabricate structures as small as 6 μm in glass, polymers, ceramics, and metals. Depending on the settings of the laser, the properties of glass can be modified without actually removing any material [2]. This technique can be used, for example, to change the refraction index of glass and thereby fabricate optical waveguides.

## 7.4
## Powder Blasting

Powder blasting has recently been introduced as a bulk micromachining technique for brittle materials, such as glass, silicon, and ceramics. This technology is based on mechanical material removal from a substrate by a jet of particles. A mask containing the design covers the substrate, and then particles are accelerated towards the target by high-pressure airflow through a nozzle. Because of their high resistance to powder blasting, metals and elastomers may serve as mask materials, and therefore standard lithographic techniques can be used to define complex designs. The removal rate by powder blasting depends on the substrate materials properties, such as the Young's modulus, hardness, and fracture toughness, and on the kinetic energy of the powder particles. As reported by Wensink and Elwenspoek [3], removal rates of 25 $\mu m$ $min^{-1}$ can be achieved for glass and silicon. The minimum attainable feature size with powder blasting is about 30 $\mu m$ with aspect ratios up to 2.5 [4]. Like wet etching, powder blasting is not a selective fabrication method. Additionally, this technique creates rough surfaces, which makes powder blasting not particularly attractive for the fabrication of $\mu TAS$.

## 7.5
## Glass Bonding

For sealing a glass substrate with an etched microfluidic channel network, another glass substrate is often preferred as the cover plate. This type of device has been the workhorse in capillary electrophoresis experiments for a decade (section 10.4). The three most frequently used glass-to-glass bonding methods are thermal fusion bonding, anodic bonding, and adhesion bonding. Among these methods, fusion bonding has been the most popular, because it is a direct bonding method, which means that no intermediate layers are necessary. An additional reason for its popularity is that the only equipment needed is a furnace and an etching bath for preparation of the wafer surfaces. This is fortunate, since laboratories that do not have cleanroom facilities also can work with chemical analysis on microfabricated devices.

Even though the mechanism of fusion bonding is not completely understood, it is believed to rely on a chemical reaction between hydroxide groups (OH–) present in the interface of the wafers (Fig. 7.1) according to the following reaction:

$$Si-OH + OH-Si \quad \rightarrow \quad Si-O-Si + H_2O$$

in which water ($H_2O$) is formed and released under heating, causing the hydrogen bonds between the silanol groups (Si–OH) to turn into covalent siloxane bonds (Si–O–Si).

It is clear from Figure 7.1 that the wafer surfaces have to have exposed hydroxide groups for bonding to be successful. This is typically done by hydrophilization

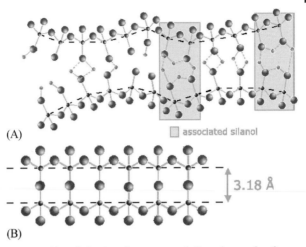

(A)

☐ associated silanol

(B)

3.18 Å

**Fig. 7.1** Glass fusion bonding process. A: Two glass wafers face each other. In some places, the surface hydroxide groups (OH⁻) form silanol bonds, which pull the glass wafer together. These initial forces facilitate the subsequent fusion bonding.
B: After the glass wafers have been heated under light pressure, silanol bonds between the glass surfaces form siloxane bonds (Si-O-Si) according to Equation 7.2 (figures courtesy Steen Weichel).

or hydration in a so-called Piranha solution, which is a mixture of $H_2O_2$ and $H_2SO_4$. This is advantageous, because a hydrophilic surface has many silanol groups. Treatment in an oxygen plasma can also be used to hydrophilize the surfaces. The wafers are typically dried and placed in a furnace after assembly. Ramping the temperature slowly to below the glass transition temperature (600–650 °C) results in bonding if the wafer surfaces are smooth enough. A load is also often placed on the wafer pair to facilitate bonding.

An alternative method that sometimes gives better results is to leave a thin film of water between the wafers. The temperature is ramped to below the boiling temperature, e.g., 80 °C, to let the water film evaporate slowly. The temperature is subsequently increased to below the glass transition temperature.

The smoothness of the wafers is critical, because sufficient contact area is necessary to achieve a successful bond, and the contact area is controlled mainly by the surface roughness. Commercially available glass substrates are polished to a roughness of about 1 nm, which is generally sufficient for fusion bonding.

Even though the bonding procedure seems straightforward, it is in fact almost black magic to get it to work, because it is so sensitive to the cleanliness and smoothness of the substrates. The quality of the bond also depends highly on the bonding area [5]; it is generally easier to bond small pieces of glass than whole wafers.

Fusion bonding of fused silica substrates is also possible; a temperature of about 1000 °C is typically needed, due to the higher glass transition temperature of this material.

Anodic (or electrostatic) bonding is a field-assisted thermal bonding technique typically used to bond a structured silicon substrate to a borosilicate glass cover plate (see Chapter 6). This bonding method generally has a higher yield than thermal fusion bonding, because the wafers are kept in contact by electrostatic forces, so the requirements concerning surface roughness and cleanliness are considerably relaxed. Two glass substrates can unfortunately not be anodically bonded without the use of at least one intermediate layer, such as polysilicon [6].

Adhesion bonding or gluing has also been widely used for assembling wafer pairs. The success of this packaging solution is determined by the adhesive surface tension (wetting), which depends on the surface energy of the material, the adhesive viscosity, and the surface cleanliness. The strength of this sealing depends on both adhesion forces and the internal strength (cohesion) of the adhesive itself. One can, for example, use polymeric adhesives that turn from the liquid to the solid state by various polymerization reactions (UV exposure, anionic reaction, in the absence of oxygen, thermal cure, etc.). A thin layer of adhesive is typically spun onto one of the wafers (e.g., the cover) in the same manner as spinning of photoresist (section 6.2.1). The wafers are pressed slightly together and the glue is hardened.

An advantage of this bonding method is that it can be done at room temperature, for example for polymerization upon UV exposure. A disadvantage is that the bonding process can affect the definition of the fluidic channels, possibly even clogging them, if they become filled with glue. Another disadvantage is that polymer adhesion layers generally have a much lower chemical resistance than glass and that they may be toxic, which is true for some UV-cured adhesives.

An alternative solution is to use liquid glasses as an adhesive material. Liquid water glasses are silicate materials (quartz sand and solid water glass crystals), dissolved in alkaline solutions (sodium or potassium), having glass characteristics. Due to their high alkaline properties, liquid glasses are soluble in water, mixing in all ratios.

## 7.6
## A Microfabrication Example

The vast majority of microdevices for capillary electrophoresis are fabricated by isotropic etching and fusion bonding of borosilicate substrates. The fluidic layout typically consists of a channel cross (Fig. 7.2).

An often-used processing recipe is shown in Fig. 7.3 as a general example. (The optimal processing conditions usually differ among different laboratories.)

First, the bottom substrate has to be cleaned, which can be done chemically, mechanically, or by a combination of chemical and mechanical cleaning. It should at least be cleaned chemically, for example by immersion into a $1:3$ mixture of $H_2O_2$ and $H_2SO_4$ for 10 min (Piranha solution). A chromium adhesion layer about 100 nm thick is deposited on the substrate (Fig. 7.3, step a), so the photoresist does not delaminate and fall off during etching. Photoresist is subsequently

**Fig. 7.2** Layout of a typical glass chip for electrophoretic separations. The channel lengths are in the centimeter range.

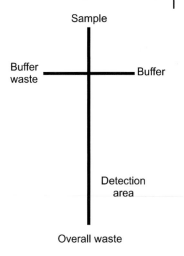

**Fig. 7.3** Schematic processing sequence for fabricating microchannels on a bottom substrate. See text for detailed description.

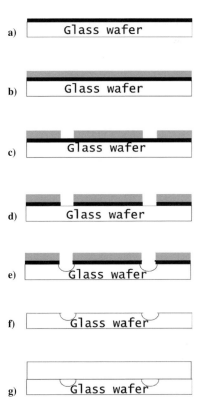

spun onto the wafer (step b), and the mask pattern is transferred to the resist by standard UV photolithography (step c, see section 6.2.1). A resist thickness of about 4 μm should be enough to etch channels to a depth of 10–20 μm. In step d) the wafer is immersed in a chromium etchant to expose the areas of the channel network. The chromium is removed within a few minutes at an etching rate of about 60 nm min$^{-1}$ when the following solution is used: 15 g cerium sulfate, 1215 mL $H_2O$, and 90 mL $HNO_3$. The channel network can be etched to a depth of 10–20 μm in 40% HF. The etching rate is about 3 μm min$^{-1}$, depending on the type of substrate (step e). The photoresist is stripped and the remaining chromium layer is removed by immersion for a couple of minutes in the chromium etchant (step f).

The inlet holes in the lid have to be fabricated before fusion bonding of the substrates (step g). A crude way of doing this is by drilling. It is, however, difficult to keep the substrate clean with this method, which is a disadvantage. The through-holes can alternatively be fabricated by etching, by the following procedure: The cover substrate is laminated with a HF-resistant polymer film. At the locations of the access holes, the polymer film is removed by, e.g., $CO_2$ laser ablation. Then the access holes are etched through the substrate with 40% HF, and the polymer film is removed.

The two substrates have to be rendered hydrophilic before fusion bonding. This can be done by immersion in a Piranha solution for 10–30 min (similar to the initial cleaning procedure before step a). The substrates are then pressed together and annealed in a furnace at 600–650 °C for a couple of hours to increase the bond strength. A more detailed description of fusion bonding can be found in section 7.5.

## 7.7
## References

1  I. FANDERLIK *Optical Properties of Glass.* Elsevier Science Ltd., Amsterdam, **1989**.

2  P. BADO *Laser Focus World,* **2000**, 36(4).

3  H. WENSINK and M.C. ELWENSPOEK, *Sens. Actuators, A,* **2002**, 102, 157–164.

4  E. BELLOY, A.-G. PAWLOWSKI, A. SAYAH and M.A.M. GIJS. *J. Microelectromech. Systems,* **2002**, 11, 521–527.

5  M. STJERNSTRÖM and J. ROERAADE, *J. Micromech. Microeng.,* **1998**, 8, 33–38.

6  D.-J. LEE, Y.-H. LEE, J. JANG and B.-K. JU. *Sens. Actuators, A* **2001**, 89, 43–48.

7  H. BECKER, M. ARUNDELL, A. HARNISCH and D. HÜLSENBERG, *Sens. Actuators, B,* **2002**, 86, 271–279.

# 8
# Polymer Micromachining
Henning Klank

**Concept**

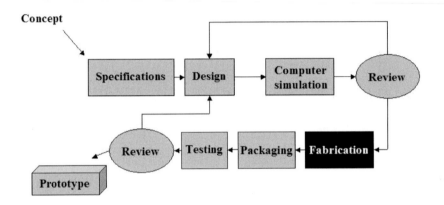

The introduction mentioned that in some cases polymers are better suited for fabricating µTAS than silicon. For example: many types of polymers show better resistance to chemical treatment and better biocompatibility. Furthermore, polymer microstructures can be mass produced at significantly lower cost than silicon microstructures. Plastics come in various forms and shapes, ranging from a flexible transparent sheet to a hard, colored block. It is often possible to find a polymer that has the optical qualities desired for a given application. For example, some polymers, such as PDMS, are transparent in the ultraviolet region of the electromagnetic spectrum, and most thermoplastic polymers, such as PMMA and PC, are transparent in the visible region. There are even polymers that are transparent in the infrared region, for instance, PC and polyetherimide.

As the name implies, polymer materials consist of one type of polymer or a mixture of polymers. Polymers consist of long chains of smaller molecules that are the basic building blocks of polymers, called monomers. Polymers are based on the element carbon, which allows the building of long-chain molecules. Bulk polymer materials are called plastics. The way polymer chains are arranged within the polymer material determines the macroscopic properties of the plastic. Plastics are classed into three major groups: elastomers, thermoplastics, and thermosetting plastics. All three types can be used in fabricating µTAS, but thermoplas-

*Microsystem Engineering of Lab-on-a-chip Devices*
O. Geschke, H. Klank, P. Telleman
Copyright © 2004 Wiley-VCH Verlag GmbH & Co. KGaA, Weinheim
ISBN: 3-527-30733-8

tics are currently the ones used most often for micromachining, because a wide range of microfabrication methods is available for them. Thermoplastics are materials in which the polymer chains are aligned with one another. Polymers of this type are often hard, brittle, and optically transparent at room temperature and are thus similar to glass and known as organic glasses. Thermoplastics that are often used include polymethylmethacrylate (PMMA), polycarbonate (PC), and cycloolefin copolymer (COC). When a thermoplastic is warmed, it transitions into a softened, rubbery state, due to the sliding of polymer chains against each other prior to melting. The temperature at which the polymer begins to soften is called the glass transition temperature. The softened state and the glass transition temperature play an important role in the machining of thermoplastics. Warming and then cooling a thermoplastic is known as thermocycling, a process that is not necessarily reversible. When cooling, the polymers might align to a greater or lesser degree, depending on the time available. If a structure is stressed as a result of mechanical cutting or drilling, thermocycling can be used to relieve the stress. This process is known as annealing (see also section 6.6.3).

The two major ways of machining polymers are replication from a master and direct machining. Replication methods often produce a microstructure by allowing a polymer workpiece to form an inverse copy of a mold. At least three different methods can be used to make such a copy: hot embossing, injection molding, and casting. These methods have in common that a very soft or even liquid form of polymer is poured or pressed into a mold, after which the material is hardened and removed from the mold. Direct machining methods remove small amounts of polymer in places where microstructures, such as microchannels or microwells, should be located. The material is removed either mechanically, e.g., by a computer-controlled microdrill or sandblaster, or by means of radiation, e.g., irradiation with infrared or ultraviolet light or x-rays. In comparing different polymer machining methods, we can consider various quality and quantity related factors (Lee et al. 2001). The most important factors that come to mind are cost, minimum feature size, production speed, and complexity of the process. All these are discussed in the following sections.

## 8.1
## Hot Embossing

Hot embossing (Fig. 8.1) is the process of pressing a mold into a heated, softened thermoplastic, followed by cooling, thereby producing an inverted replicate of the mold (Fig. 8.2). The first step in hot embossing consists of heating the mold and the thermoplastic to the glass transition temperature. Once the thermoplastic begins to soften it takes on the form and shape of the mold. The mold and thermoplastic are then cooled below the glass transition temperature to harden the thermoplastic. Finally, the thermoplastic is removed from the mold. The mold can be reused many times, depending on its mechanical strength. The mold can be produced of any kind of material that can sustain the elevated temperatures and pressures used in hot embossing. Both silicon and metals are used as molds in the

**Fig. 8.1** Hot embossing. Step 1, a mold and a thermoplastic polymer workpiece are arranged above each other. Step 2, the mold and workpiece are clamped together and heated. Step 3, workpiece and mold are cooled and separated.

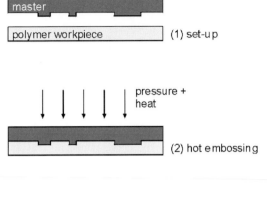

(1) set-up

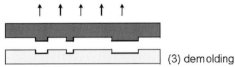

pressure + heat

(2) hot embossing

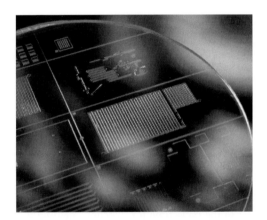

(3) demolding

**Fig. 8.2** Hot embossed plastic structures.

hot embossing of plastic microstructures. The hot embossing process of course depends strongly on the chosen thermoplastic, because different thermoplastics have different flow properties when they are soft, different thermal expansion coefficients, and different glass transition temperatures. Each type of polymer needs its own temperature and pressure settings to optimize the hot embossing process.

Hot embossing can be used to create microstructures with feature sizes smaller than 100 nm (D'Amore 2000). The ability to faithfully reproduce structures at the nanoscale level has led to the term nanoimprinting (Chou et al. 1995). The driving force behind nanoimprinting is the idea of being able to supplement lithogra-

**Fig. 8.3** Injection-molded structures.

phy techniques with hot embossing to fabricate large numbers of nanostructures in plastics.

## 8.2
## Injection Molding

The injection molding technique is similar to hot embossing in many ways. Here too an inverted copy of a mold is made (Fig. 8.3). Injection molding differs from hot embossing mainly in the fact that molten polymer material is injected into the mold. For microinjection molding, the mold is heated to the softening temperature of the polymer to prevent the injected polymer material from hardening too early (Rötting et al. 2002). After injection of molten polymer into the mold, the mold is cooled slowly, so that the polymer hardens. The polymer microstructure can then be removed from the mold. The mold has to be durable enough to withstand the high pressures that are used, which means that by far most molds for injection molding are made of metal. Like hot embossing, injection molding can reproduce structures with features as small as several nanometers. Injection molding is however the more suitable method for producing many replicates, because the cycle time for injection molding is shorter than for hot embossing.

## 8.3
## Casting

Casting is a rather straightforward process, which rapidly yields microstructures. The smallest feature sizes depend mainly on the quality of the mold and can be at the nanoscale (Whitesides and Love 2001). Casting is often used in research laboratories for fast production of moderate numbers of prototypes. Hot embossing and injection molding have in common that they use a heating and cooling cycle to soften and structure the polymer on a mold. Casting uses chemical processes to harden the polymer. Two components, a base and a hardener or cure, are mixed just prior to use. Immediately upon mixing, the chemical curing process starts and, after a certain amount of time, results in hardening of the polymer. The liquid mixture is poured in the mold and as the polymer sets the polymer takes the shape of the mold. The polymer structure can then be removed from the mold.

With chemical curing, both hard and elastomeric polymer structures can be produced. For the fabrication of microstructures, elastomers are very popular because they form hermetic, reversible seals to smooth surfaces like glass and silicon by adhesion. The elastomer polydimethylsiloxane (PDMS) is the most frequently used elastomer for casting microstructures. Elastomers also adhere very well to another sheet of the same material, in which case the sealing forces are based on cohesion. The temporary seal of elastomers can be made permanent by chemical treatments such as plasma activation.

## 8.4
## Laser Micromachining

In contrast to the replication methods, laser micromachining is a direct machining method. It is based on the removal of polymer material by using intense UV or infrared radiation provided by a laser. The radiation wavelength used affects the removal mechanism. If infrared lasers are used, the irradiated material is heated and decomposes, leaving a void in the polymer material. If UV radiation is used, the irradiated polymer decomposes, presumably by a mixture of two mechanisms: thermal and direct bond breaking. Thermal bond breaking is induced by heat, as with infrared radiation. In direct bond breaking, polymer molecules directly absorb ultraviolet photons, often absorbing enough energy so that the chemical bonds within the polymer chains are broken. The resulting smaller polymer chains are volatile or melt at much lower temperatures than the bulk polymers, thereby leaving a void in the material.

Fig. 8.4 shows a laser micromachining system for direct writing, which can produce microstructures by driving a focused laser beam across a polymer workpiece. The microstructure is designed with the help of a 2D computer-aided design (CAD) program. The computer is connected to a marking head that steers the laser beam in two orthogonal directions. The laser beam is focused on the workpiece, which is positioned in a sample holder. Translation of the coordinates provided by the software into movements of the laser beam results in the formation of microchannels and cavities. The intensity of the laser beam can be varied simultaneously with the beam velocity and the number of passes over one particular microstructural detail. The various elements of a microstructure are written sequentially, which requires optimization of the writing sequences as to achieve the best results. The movements of the laser beam across the polymer workpiece can be 100–1000 mm s$^{-1}$, allowing fabrication of microstructures within seconds.

The smallest feature size attainable with laser machining depends strongly on the quality of the optical system and the laser wavelength. The relation between

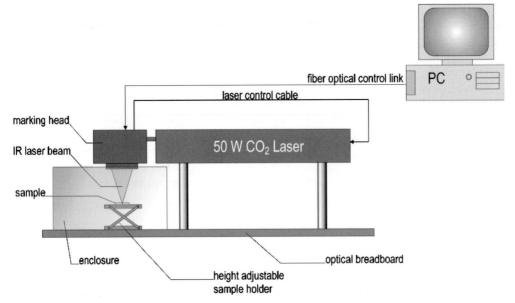

**Fig. 8.4** Laser micromachining system consisting of a laser, a laser marking head, a personal computer (PC), a power supply, and a sample holder, including a sample and an enclosure. The laser beam enters the marking head, where the beam direction can be manipulated under computer control. The beam is then focused onto the sample. Any 2D microchannel or well network can be designed on the personal computer and translated into corresponding movements of the focused laser beam.

the smallest possible focal spot diameter $d$, the relative aperture of the optical system $a$, and the wavelength $\lambda$ is

$$d = \frac{\pi}{4} \frac{\lambda}{a} \qquad (8.1)$$

where the relative aperture $a$ is defined as the ratio between the aperture of the optical system $D$ and the focal length of the focusing lens $f$:

$$a = \frac{D}{f} \qquad (8.2)$$

We can conclude from Eq. 8.1 that for small feature sizes, short-wavelength radiation should be used together with an optical system of high relative aperture. This immediately shows the advantage of ultraviolet over infrared radiation. In practical terms the feature sizes that can be obtained with an infrared laser micromachining system are typically around 200 μm, whereas the spot size for an ultraviolet system can be as small as 1.5 μm (Becker and Locascio 2002).

Ultraviolet laser machining systems can be operated as direct writing systems, but systems that work with masks are much more common. Here, the laser beam is projected onto the workpiece through a mask instead of manipulating the position of a focused beam. Feature sizes thus depend on the quality of the mask rather than on the wavelenght of the laser.

## 8.5
## Milling

Milling or micromilling is a mechanical method that can be used to produce polymer microstructures. For micromilling, a small revolving cutting tool mechanically removes polymer material. The setup that is used for micromilling looks very much like the laser machining setup. A computer controls the position and movement of the cutting tool. This process is known as computer numerical control or CNC milling. Like laser machining, CNC milling is a serial process. CNC milling is, however, considerably slower than laser machining, partly because cutting tools tend to be fragile and break easily. The production time for a microstructure lies anywhere between a few minutes and half an hour, depending on the number of details within the microstructure. Although CNC milling cannot achieve the very small feature sizes of the replication techniques it can produce structures with sizes down to 100 µm. It is relatively straightforward to operate a CNC milling machine, and the milling process is very flexible. In comparison to laser micromachining, milling has the advantage that the polymer workpiece material is not chemically degraded by heating or UV radiation. Milling can leave tensions and stress near a cut groove or cavity, which can be relieved by annealing the structure by a cycle of heating and slow cooling. Micromilling can only be used with hard materials that are easy to cut. Thermoplastics are usually good milling materials, as long as the milling does not warm the material too much. Elastomers, on the other hand, usually cannot be milled.

## 8.6
## X-ray and Ultraviolet Polymer Lithography

Polymer lithography is based on the fact that some polymer materials are affected by energetic radiation such as UV radiation. The radiation may break chemical bonds or lead to other types of chemical change within the polymer material. Lithography is a mask-based technique, described in detail in section 6.2.2. The purpose of the mask is to absorb or reflect the radiation in certain places while allowing radiation to arrive at places where microchannels and microwells should be. The most widespread method of polymer lithography is the fabrication of microstructures with photoresists, which are discussed in detail in section 6.2. One of the most-often used photoresists is SU-8, which enables the creation of films up to about a millimeter thick. Irradiated surfaces of SU-8 are cross-linked and made insoluble. Areas of SU-8 that are not irradiated can be removed by developing, leaving the microstructured SU-8 photoresist standing. Structures with high aspect ratios of up to 20:1 are producible. The feature size achievable with UV lithography is about 1 µm. The complete photolithographic process including applying the photoresist, preparing it for irradiation, and developing does not take more than an hour.

X-rays can also be used to structure polymers. The LIGA (from the German Lithographie, Galvanoformung und Abformung, i.e., lithography, electroplating, and molding) process was developed in the early 1980s and uses x-ray irradiation through a thick gold mask to structure the plastic polymethylmethacrylate (PMMA). The chemical structure of PMMA is altered by irradiation with x-rays, and it can then be removed with an appropriate solvent. In LIGA the resulting polymer microstructure is an intermediary step in making a metal mold by means of electroplating (section 6.3).

## 8.7
## Sealing of Polymer Microstructures

To enable fluids to be manipulated within microstructures, such as in chips used for chemical or biochemical assays, it is essential that the structures are hermetically sealed. Some possibilities for sealing a microstructure include gluing, laminating, direct and thermal bonding, and ultrasonic and laser welding.

One of the best known methods for joining two plastic surfaces is gluing, in which a thin film of glue is applied to one or both surfaces. The two parts are then pressed together. The glue, which is often a polymer itself, hardens to form a bond between the two polymer parts. The biggest issue when considering gluing microsystems is the fact that the glue might flow or creep into the microstructures, thereby destroying their functions. Especially when working with microstructures containing microchannels, it is difficult to prevent the glue from flowing into the channels by capillary action. Gluing is therefore not often the best solution for bonding two microstructured polymer parts. Laminating resembles gluing in many ways. A thin sheet of lamination foil is heated and pressed onto the microstructured polymer workpiece. The lamination foil contains a thin layer of glue that forms a bond to the workpiece after both the foil and workpiece are cooled. Lamination suffers from the same drawbacks as gluing and has only limited use in the sealing of microstructures.

Direct bonding is the adhesion or cohesion of one flat piece of material to another (Spierings et al. 1995). Adhesion and cohesion are based on a combination of the many physical and chemical bonds that are established when surface molecules of the two joining parts are brought very close together. For hard materials such as glass or silicon, it is difficult, although not impossible, to achieve the necessary surface smoothness to enable the surface atoms to come close enough together. This is why anodic bonding is preferred for bonding silicon to glass (see also section 6.6.4). Elastomers such as PDMS can assume curved profiles and thereby facilitate direct bonding. Direct bonding does not create strong bonds between the workpiece and cover and so is temporary. Direct bonding is thus very useful in applications where the lid of a microstructure must be removed to access a certain part of the microstructure.

Thermal bonding is a process in which the parts to be bonded are pressed together at elevated temperatures. Spierings et al. (1995) believe that the thermal bonding process is due to melting of the surface layers of the polymer workpieces, followed by intertwining of the polymer chains of the two plastic pieces that are pressed together. After cooling, the two parts are permanently bonded. Obviously, using elevated temperatures results in softening the polymer and may cause destruction of the microstructures elements, especially under pressure. Care must be taken to find the right pressure and temperature to allow efficient bonding without compromising the integrity of the microstructure. Under optimized conditions, thermal bonding is excellent for bonding microstructures.

Laser welding is a localized thermal bonding process, inasmuch as the interface between workpiece and lid is briefly melted and then cooled again. Local heating is usually achieved by lasers. Laser welding requires the workpiece to absorb laser energy and the lid to be transparent. For bonding, laser light is directed through the lid to the workpiece. As the laser light is absorbed in the workpiece it locally heats and melts the workpiece and lid together. After cooling, the workpiece and lid are permanently bonded. To assure that the workpiece absorbs sufficient energy from the laser, it is usually colored, which may be undesirable in certain applications. Laser welding does not therefore represent a generic bonding method for microstructures.

## 8.8
## Adding Functionalities

Similar to silicon microstructures, the possibility of adding functionality to polymer microstructures should be highly beneficial. We next discuss the integration of metal electrodes and waveguides into polymer structures, because both are essential elements in many detection schemes in chemical and biochemical analyses.

The most important examples of integrated metal electrodes are microfluidic systems that use electroosmotic flow (section 4.1) and systems that use electrochemical detection (section 4.6). Electrodes can be added to polymer microstructures by adding bulk metal electrodes, for example platinum wires in reservoirs in a microstructure to drive electroosmotic flow. However, in many applications the electrodes must be embedded in the microstructure. Thin film deposition of metals (section 6.3) on polymers is therefore preferable. For plastic microsystems, shadow masking is a promising technique, because it does not require the polymer material to be set to a certain electrical potential. The metal that is intended for the electrode material can be evaporated or sputtered onto the polymer workpiece through a mask. Other methods of metallizing polymer surfaces are physical vapor deposition and plasma sputtering (section 6.3.2). Not all metals are equally easily applied to all polymer materials. Problems that are often observed include the formation of metal clusters instead of uniform films probably due to stress in the metal film (Ohring 1992), alteration of the chemical structure of the polymer under the thin-film deposition conditions, and formation of microcracks in the metal film due to differences in thermal expansion coefficients between metal and polymer. Polymers can also be coated with other inorganic materials, e.g., silicon or aluminum oxide (Benmalek and Dunlop 1995).

Another important functionality that plays a central role in the application of microsystems in chemistry and biochemistry is optics. Optical sensors are discussed in detail in section 4.5, where integrated planar waveguides were introduced. Waveguide structures have traditionally been fabricated in glass. In recent years however, polymer waveguides have been fabricated. One example relies on structuring a single layer of the photoresist SU-8 for defining optical components such as planar waveguides. SU-8 has a higher refractive index than air, and light that is coupled into a SU-8 waveguide remains in the wave guide. SU-8 waveguides can be combined with channel networks to enable absorption and fluorescence measurement on chip.

Surface modification of polymers enables local changes in the properties of the polymer and thereby increased functionality. A very important example of surface modification is the treatment of polymers to establish local hydrophobic or hydrophilic areas, which can be used as burst valves to direct flows on a chip (Fig. 4.7). Areas can be added to a chip that allow binding of biomolecules like DNA or proteins. Or specific areas of the chip can be treated to prevent binding of biomolecules. Polymer surfaces are usually modified by plasma treatment or wet chemistry. One example of introducing charges onto polymer surfaces is oxidization with an oxygen plasma (Duffy et al. 1998). Oxidization leads to a build-up of a negative charge that, for example, enables electroosmotic pumping in the polymer microfluidic system. Surface charges can also be added by UV laser photoablation to produce carboxylate surface groups (Johnson et al. 2001). Many methods for surface modification of polymers exist; their success depends strongly on the properties of the polymer. Care must be taken to test different techniques to find the one that yields the best results.

The ability to integrate functional components into polymer microstructures is certainly not limited to metal electrodes, optical components, or surface modifications. In contrast to the integration of functional components on silicon chips, however, integration of functional elements onto polymer chips is rather limited. A great deal of research is being conducted on functional polymers, e.g., conducting polymers, which may have interesting applications in microtechnology applied to chemistry and the life sciences. We will undoubtedly see increasing efforts in developing polymer micromachining to the same level as silicon micromachining.

**8.9**
**Examples of Polymer Microstructures**

There are a few good examples of polymer microfluidic systems. This section presents two such examples of systems that are at least partly functioning or prototypes.

One example of a commercial polymer microsystem is produced by the Swedish company Gyros (Ehrnström 2002, section 4.1). The Gyros system is made in the same way that audio compact discs are produced, by injection molding of a polymer wafer on a metal mold. After injection molding, the polymer wafers are further processed to define valves, modify surfaces to achieve compatibility with biochemical assays, and finally to bond a lid onto the wafer.

The Gyros disc features microchannels that are 100 μm wide or wider. Within a so-called Gyrolab microlaboratory, applications can be miniaturized and integrated into single streamlined procedures. Each CD-like microlaboratory can contain hundreds of identical application-specific microstructures (e.g., Fig. 8.6).

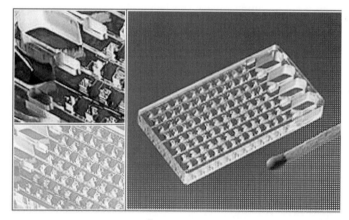

**Fig. 8.5** Photograph of Lilliput® microtiter plate fabricated by Steag Microparts, Germany. The microtiter plate is used for bacteria identification and antibiotic susceptibility tests. The plate features 96 reaction wells with a volume of 1.8 μL each. Four inoculation inlet cavities allow simultaneous, even filling of 24 wells via a microfluidic network. The microsystem is fabricated by injection molding from polystyrene (PS) polymer (figure courtesy STEAG Microparts and Merlin Diagnostika).

**Fig. 8.6** A laboratory compact disk from Gyros AB, Sweden, produced by injection molding. The microfluidic structures on the disk have widths of 100 μm and varying depths in the range 10–150 μm. The disk is made of polyolefin (figure courtesy Gyros AB).

**Fig. 8.7** Microammonia meter for detecting ammonia in aqueous solution. The system was produced with a infrared laser system, and microchannel widths and depths are 250 µm and 100 µm. The system has four inlets, for three reagents plus the sample, and one waste outlet. The reaction to detect ammonia is the Berthelot reaction. The system is made of PMMA.

Another example of a commercial microfluidic system is the Lilliput® microtiter plate produced by Steag Microparts in Germany (Fig. 8.5). The microtiter plate is an example of a system in which capillary forces are used for priming. The microtiter plate consists of 96 wells in all, of which 24 each are filled simultaneously from a single filling cavity. The Lilliput® system is produced by injection molding. The typical feature size of the microfluidic elements is $30 \times 30$ µm$^2$.

An example of a polymer microsystem that exploits favorable material properties of plastics is the microammonia meter shown in Fig. 8.7. A chemical reaction involving three steps that converts ammonia to an indophenol dye is performed in the microsystem. The change in absorption of red light by the blue indophenol dye is an indirect measure of the concentration of ammonia in a water sample (Daridon et al. 2001). The entire chemical analytical procedure is automated and therefore allows for continuous, inline chemical analysis. This structure was developed specifically for inline monitoring of ammonia concentrations in waste water in a sewage-treatment plant. The initial design and fabrication aimed at creating this structure in silicon. However, the conversion of ammonia to indophenol dye occurs at high pH, and, despite extensive efforts to protect the silicon from etching, it appeared impossible to prevent corrosion of the silicon chip. The polymer polymethylmethacrylate (PMMA) was chosen to replace the silicon chip, and the ammonia analysis results obtainable with the polymer chip look very promising.

## 8.10
## References

J.C. Ball, D.L. Scott, J.K. Lumpp, S. Daunert, J. Wang and L.G. Bachas, *Analytical Chemistry*, **2000**, 72, 497–501

H. Becker and U. Heim, *Sensors and Actuators A*, **2000**, 83, 130–135

H. Becker and L.E. Locascio, *Talanta*, **2002**, 56, 267–287

M. Benmalek and H.M. Dunlop, *Surface and Coatings Technology*, **1995**, 76–77, 821–826

P.G. Berrie and F.N. Birkett, *Optics and Lasers in Engineering*, **1980**, 1, 107–129

S. Chou, P.R. Krauss, and P.J. Renstrom, *Applied Physics Letters*, **1995**, 67, 3114–3116

A. D'Amore, D. Simoneta, W. Kaiser, H. Schift, and M. Gabriel, *Kunststoffe*, **2000**, 6

A. Daridon, M. Sequeira, G. Pennarun-Thomas, H. Dirac, J.P. Krog, P. Gravesen, J. Lichtenberg, D. Diamond, E. Ver-

POORTE and N.F. DE ROOIJ, *Sensors and Actuators B*, **2001**, 76, 235–243

D.C. DUFFY, J. COOPER MCDONALD, O.J.A. SCHUELLER and G.M. WHITESIDES, *Analytical Chemistry*, **1998**, 70, 4974–4984

A. GERLACH, G. KNEBEL, A.E. GUBER, M. HECKELE, D. HERRMANN, A. MUSLIJA, TII. SCHALLER, *Microsystem Technologies*, **2002**, 7, 265–268

M. GITIN, *Photonics Spectra*, **1998**, 9, 136–139

M. HECKELE, W. BACHER and K.D. MÜLLER, *Microsystem Technologies*, **1998**, 4, 122–124

T.J. JOHNSON, E.A. WADDELL, G.W. KRAMER and L. LOCASCIO, *Applied Surface Science*, **2001**, 181, 149–159

H. KAWAZUMI, A. TASHIRO, K. OGINO and H. MAEDA, *Lab on a Chip*, **2002**, 2, 8–10

L.J. LEE, M.J. MADOU, K.W. KOELLING, S. DAUNERT, S. LAI, C.G. KOH, Y.-J. JUANG, Y. LU and L. YU, *Biomedical Microdevices*, **2001**, 3–4, 339–351

L. MARTYNOVA, L. LOCASIO, M. GAITAN, G.W. KRAMER, R.G. CHRISTENSEN and W.A. MACCREHAN, *Analytical Chemistry*, **1997**, 69, 4783–4789

R.M. MCCORMICK, R.J. NELSON, M.G. ALONSO-AMIGO, D.J. BENVEGNU and H.H. HOO-PER, *Analytical Chemistry*, **1997**, 69, 2626–2630

R.F. MIRACKY, *Laser Focus World*, **1991**, 5, 85–98

M. OHRING, *The Materials Science of Thin Films*, **1992**, Academic Press, San Diego.

J. POWELL, *CO$_2$ Laser Cutting*, Springer, Berlin, **1998**

M.A. ROBERTS, J.S. ROSSIER, P. BERCIER and H. GIRAULT, *Analytical Chemistry*, **1997**, 69, 2035–2042

O. RÖTTING, W. RÖPKE, H. BECKER and C. GÄRTNER, *Microsystem Technologies*, **2002**, 8, 32–36

S. SCHLAUTMANN, H. WENSINK, R. SCHASFOORT, M. ELWENSPOEK and A. VAN DEN BERG, *Journal of Micromechanics and Microengineering*, **2001**, 11, 386–389

G.A.C.M. SPIERINGS, J. HAISMA, and F.J.H.M. VAN DER KRUIS, *Philips Journal of Research*, **1995**, 49, 139–149

R. SRINIVASAN, *Journal of Applied Physics*, **1993**, 73, 2743–2750

G.M. WHITESIDES and J.C. LOVE, *Scientific American*, September **2001**, 33–41

# 9
# Packaging of Microsystems

Gerardo Perozziello

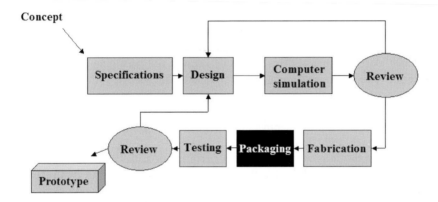

For microsystems to function efficiently, it is vital to package them carefully. Detailed knowledge of materials, device behavior, and reliability, as well as an understanding of the limits of current packaging technology are required for successful packaging.

Critical, challenging packaging issues include microfluidic interconnects (connecting device to device and interfacing macrocomponent to microdevice), providing electrical connections while maintaining insulation between electronics and fluids, optical interconnections, and mounting and encasing devices that provide an appropriate interface to other devices or to the external environment.

The functions of packaging are to protect the devices from the environment and also to protect the environment from the device operation. Protection of the device includes electrical insulation and passivation of leads and device structures from penetration of moisture and ions. Sealing techniques, hermeticity measurements, and mechanical protection are used to ensure structural integrity and dimensional stability, thermal and optical isolation, and chemical and biological protection.

It is also necessary to protect the environment from the device materials and the device's operation, so that no undesirable reaction with the environment or contamination of it occurs. This is especially important for devices used in bio-

*Microsystem Engineering of Lab-on-a-chip Devices*
O. Geschke, H. Klank, P. Telleman
Copyright © 2004 Wiley-VCH Verlag GmbH & Co. KGaA, Weinheim
ISBN: 3-527-30733-8

medical, pharmaceutical, and food-processing applications. Biocompatibility and contamination must be considered as factors in the design. For example, chemical and biological transducers interact directly with solids, gases, and liquids of all types. In this sense, they require 'vias' or inlets in their packaging to permit this interaction (also required for some other sensor types, such as pressure sensors).

In IC technologies, packaging provides the following main functions: signal redistribution, mechanical support, power distribution, and thermal management. Signal redistribution provides more space to accommodate the interconnection capacity of a traditional wiring board for electrical contacts that are too closely spaced. The contacts are redistributed over a larger, more manageable surface by the packaging. Rigidity, stress release, and protection from the environment (for instance, against electromagnetic interference) are ensured by the mechanical support. Power distribution and thermal management should ensure a low working temperature to sustain operation for the product lifetime.

Packaging of micromachines is significantly more complex than packaging of integrated electrical circuits. This is mainly because the packaging has to protect the device from the environment and has to enable interaction with it in order to measure or influence the desired physical or chemical parameters. Devise protection has developed additional problems with the advent of integrated sensors. Silicon circuitry is sensitive to temperature, moisture, magnetic fields, electromagnetic interference, and light, just to name a few.

The package has to protect the on-board circuitry but at the same time enable it to make its measurements during the exposure. For example, in an in vivo integrated sensor, a true hermetic seal is necessary to protect the circuitry from the blood.

In addition, communication links, cooling/heating, and a means for handling and testing should be provided by the package. Also the choice of the material is very important in packaging, because the material contributes to physical protection against normal handling during and after assembly, testing, and mechanical shock. During assembly and in practical use, the package needs chemical protection. For instance, the packaging needs a series of cleaning steps during the process of installing dies on substrates and, while the device works, the package may be exposed to oxygen, moisture, oil, gasoline, and salinity. Finally, the interior environment has to be compatible with the device performance and reliability and the package should also provide this. For example, a high-quality resonator (high-Q resonator) might need a good vacuum.

The packaging and interconnection problem must be investigated in the design phase. It is necessary to solve these problems in the early design phase, because the packaging introduces physical limits to the design and construction of the system.

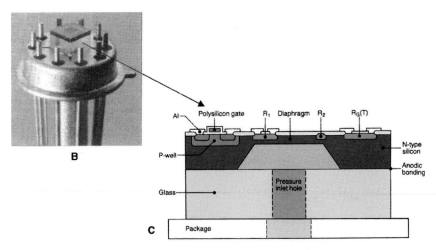

**Fig. 9.1** Pressure sensor from NEC mounted on a TO header (Madou).

## 9.1
## Levels of Packaging

Packaging of microsystems involves additive and subtractive processes, bonding, wafer scribing, lead attachment, encapsulation in a protective body, and interconnections with other devices and the environment.

Packaging can be divided into three major groups: wafer-level packaging, multichip packaging, and nonstandard packaging. Wafer-level packaging is related to the packaging of a single chip fabricated with conventional microtechnologies; multichip packaging is related to the need of packaging more chips together; and nonstandard packaging comprises other package types dictated by the specific application. For nonstandard packaging, sometimes none of the conventional microfabrication methods can be used.

The various levels of packaging and their interconnections are illustrated in the following figures.

In Fig. 9.1 the single pressure sensor belongs to the first level of packaging, and the sensor mounted on the TO header represents the third level of packaging.

### 9.1.1
### Wafer Level Packaging

Usually microdevices are fabricated by making use of wafer bonding or cavity sealing. Wafer bonding and cavity sealing can serve as 'batch'-compatible packaging techniques by encapsulating a die feature or a whole die.

Chen et al. (2002) developed a transferred ultrathin chip for micropackaging device-scale UTSi® (ultra-thin silicon is a technology used to form silicon layers on sapphire substrates without defects) by flip-chip assembly and tether-broken techniques (Fig. 9.2).

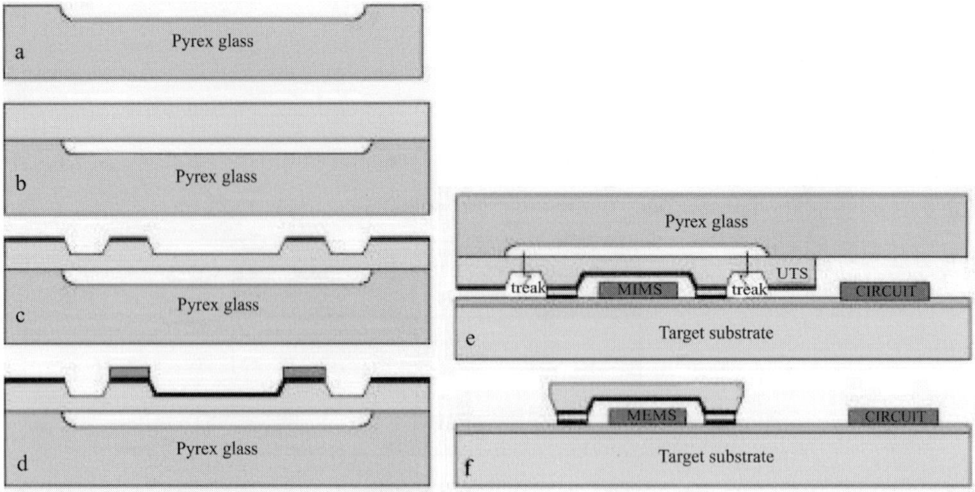

**Fig. 9.2** Process of device-scale UTSi micro-packaging: (a) BOE etches a pit on Pyrex, (b) UTSi and glass anodic bonding, (c) capping structure KOH etching, (d) solders electro-plating and reactive ion etching, (e) alignment and bonding to a host device/substrate, and (f) separation of UTSi microcaps and glass carrier (Chen et al. 2002).

Monolithic horizontal integration, in which all the components are fabricated on the same substrate in one process, takes full advantage of batch fabrication techniques, reducing the number of manual assembly steps and also resulting in small system size.

Bonded silicon and glass layers are typically rather thick and not really good for die feature-level packaging. Instead, in poly-Si and selective epitaxy surface micromachining, sealing of cavities is an integrated part of the whole fabrication process, and both die and die features can be packaged by this method. These and other lithography-defined packages, such as those involving ultraviolet patternable polymers, are used for inexpensive batch solutions.

When incorporating an intermediate layer between two substrates, often thermal bonding techniques are used (Fig. 9.3).

### 9.1.2
### Multichip Packages

Microchips can be mounted laterally as in multichip modules (MCMs) or can be stacked by using micromachining. Micromachining enables individual dies, blocks of dies, or entire wafers to be packed on top of each other, and interconnections can be run vertically from plane to plane (Fig. 9.4).

The MCM approach usually involves bonding of unpackaged chips to a substrate that provides various kinds of connections. Channels are buried within the substrate, electronic traces are patterned on the substrate surface by PC-board manufacturing techniques, and the fabricated fluidic components are surface mounted. Individual microfabricated components can be stacked or otherwise grouped to form modular systems with the advantages of flexibility, serviceability, and the ability to test parts individually before assembly.

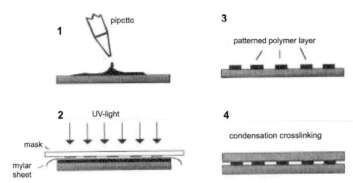

**Fig. 9.3** Procedure for silicon/polymer/silicon bonding: (1) the monomer solution is deposited by spin coating or casting; (2) the sandwich is exposed to ultraviolet light for photopolymerization; (3) the polymeric pattern is developed in xylene; (4) the second wafer is pressed onto the polymeric pattern and left for humidity-induced polymerization in air (10 h, Arquint et al., 1995).

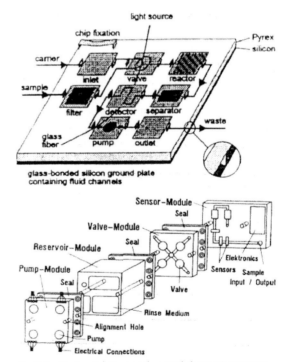

**Fig. 9.4** Horizontal (top) and vertical (bottom) concepts of microanalytical systems (Schomburg et al., 1995).

**Fig. 9.5** Micro-slit from CSEM (Swiss Center for Electronics and Microtechnology).

### 9.1.3
### Nonstandard Packages

Micro components such as pumps, flow meters, mixers, and sensors typically require device-specific, nonstandard fabrication processes, usually necessitating a hybrid package for some aspects of the system. The specific application often dictates the package.

The diverse natures of micromachining applications and the nonstandardness of the package require a design approach starting from the package. An excellent additional example of nontraditional packaging is the micro-slit, which is often used in spectrometers (Fig. 9.5). Such a slit combines silicon micromachined parts with conventional manufactured parts.

### 9.2
### Design Process in Packaging

This section describes the decision-making processes that a designer should use in formulating plans for the physical implementation of microsystem packaging.

### 9.2.1
### Phases of Design

The total design process can often be outlined as in Fig. 9.6.

The process begins with the recognition of a need and a decision to do something about it. After several iterations, the process ends with the presentation of plans to meet this need. Below, these steps in the design process are described in more detail.

**Fig. 9.6** The phases of design.

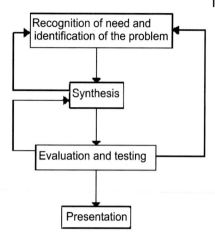

## 9.2.2
### Recognition and Identification

Sometimes the requirements of packaging are obvious. For example, the necessity of placing a pacemaker inside the human body creates means that the package must be biocompatible. The definition of the problem must also include all specifications for the item that needs to be designed.

The package specifications define the cost, the number to be manufactured, the expected life time, the range, the operating temperature, and the reliability. Moreover, the package specifications can be understood in terms of a mechanical concept with dimensions, constraints on material selection, cabling and/or interconnect requirements, thermal, mechanical and chemical environmental requirements, and finally, restrictions on manufacturing or assembly. The available manufacturing processes constitute restrictions on the designer's freedom and hence are a part of the implied specifications.

We can consider the problem of designing a package as a black box. We need to specify the inputs and outputs of the box together with their characteristics and limitations (Fig. 9.7).

The device to be packaged defines the input. For instance, if the device consists of a microfluidic system we know for sure that it needs microfluidic interconnections. The needs become the characteristics of the package, but with them we have to consider several limitations. After all these considerations we can contemplate the solutions.

## 9.2.3
### Synthesis

Having defined the problem and obtained a set of implied specifications, the next step in design is the synthesis of an optimum solution (Fig. 9.6). This phase con-

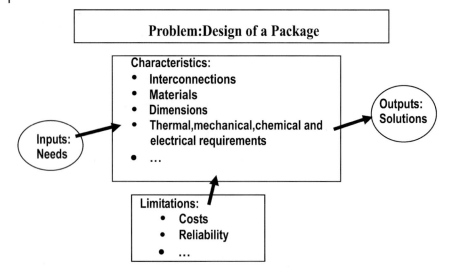

**Fig. 9.7** Symbolic representation of the definition of the problem.

sists of selecting a package fabrication process and appropriate materials, creating the corresponding artwork or drawings for the package, and writing the detailed specifications of an assembly procedure.

Now, before the synthesis, a system must be analyzed to determine whether the performance complies with the specifications. The analysis may reveal that the system is not an optimal one and that a new one has to be made. The designer should always consider mechanical, electrical, hydraulic, magnetic, electronic, and chemical properties. A problem is not solved until all likely methods of solution have been exhausted. At this point the microsystem designer should study all the solutions that can be found for each method. During this study, sketches should be as simple as possible, without adding excessive or complex detail. The aim is to obtain as many different solutions as possible, no matter how absurd they may appear at first glance.

In searching for a solution, the designer could change the normal position or character of things: if the system operates horizontally, try to operate it vertically; if it is round, try to make it square; and so on.

Fig. 9.8, for example, shows two solutions for fluidically connecting a microsystem. In one solution the external tube is glued onto the device, in the other the tube is clamped to the device by screws. Each solution is different, with different advantages and disadvantages, but the result is the same: both connect the device to fluid inlets and outlets.

The microsystem designer has to make a complete list of all the disadvantages of each proposed method of design. These disadvantages should be written down in detail and discussed. In this way the designer acts as a 'devil's advocate', looking for the faults in each design. Additionally, discussing the design with other persons helps to find better solutions or modifications to the design.

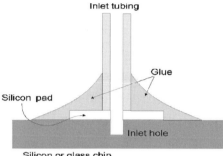

Inlet tubing

Glue

Silicon pad

Inlet hole

Silicon or glass chip

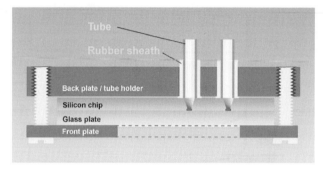

Tube

Rubber sheath

Back plate / tube holder

Silicon chip

Glass plate

Front plate

**Fig. 9.8** Schemes of two solutions for fluidic interconnections (Larsen, 2000).

Finally, the designer should make a detailed sketch of the apparently best solution for the problem. Sometimes this helps or solves at least some of the problems. Before the solution is completed there are usually a great number of changes. Having struggled with the problem and eliminated many unsatisfactory approaches, the microsystem designer has found what seems to be the best solution. The basic shape and the elements of the design are beginning to acquire a form and a substance in the designer's mind. The designer is now ready to begin the dimensional synthesis. Dimensional synthesis usually begins with a consideration of those dimensions that define factors required by the specifications.

The final phase of dimensional synthesis is related to the manufacturing and assembly problems that may be encountered.

Defining and solving a problem can be very difficult; often none of the possible approaches is satisfactory and new ones cannot be found. Only creativity permits finding the best idea to solve the problem.

9.2.4
**Evaluation and Testing**

Evaluation, which usually involves testing a prototype in the laboratory, is the final part of a successful microsystem design. Testing needs to discover if the de-

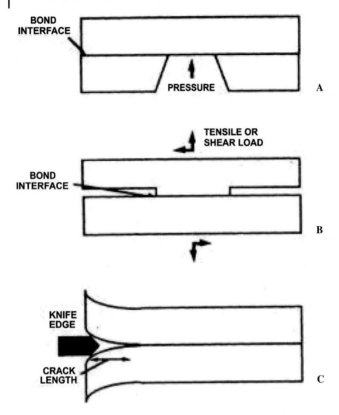

**Fig. 9.9** Three methods for bond strength measurements. (A) Burst test, (B) tensile and shear test, (C) Maszera test.

sign really meets the needs and to verify whether the system is reliable, whether it competes successfully with similar products, whether it is economical to manufacture and use, whether it is easily maintained and adjusted, and whether it is profitable. The person in charge needs to identify systems acceptance tests, package or package-component acceptance tests, calibration and trim procedures, and final acceptance test procedures. Details of the design may depend critically on the nature of the acceptance test or calibration procedure.

### 9.2.4.1 Bond Strength Test
Bond strength is investigated with the following mechanical techniques (Fig. 9.9).

The burst test (Fig. 9.9a) and the tensile/shear test (Fig. 9.9b) are used to achieve important engineering understanding of the construction of microsystems, but they do not give information about the detailed nature of the bond, because of the complicated loading of the interface. In the Maszera method a thin blade is inserted between the bonded wafers and a crack is introduced (Fig. 9.9c).

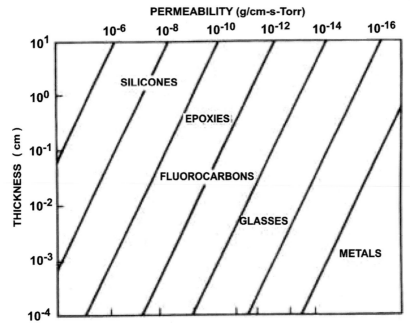

**Fig. 9.10** The calculated time for moisture to permeate various sealant materials (to 50% of the exterior humidity) in one defined geometry.

The length of the crack, calculated from infrared imaging, gives a value related to the surface energy of the sample and the elastic properties of the material (for instance a glass to silicon interface). This method has the advantage of creating a well-defined loading on the bonded interface.

### 9.2.4.2 Package Hermeticity Tests

For physical protection purposes and, in some cases, for the performance of the system inside, the hermeticity of sealed cavities has to be considered. Testing the hermeticity of several kinds of devices is usually carried out by helium leak detection. Hermeticity tests should be used wherever there are fluidic interconnections.

Fig. 9.10 shows the relative abilities of several materials to exclude moisture from encapsulated components over long period of time. Organics are orders of magnitude more permeable than materials typically used for hermetic seals (from Striny, 1988).

For measuring moisture penetration in a package, temperature-accelerated soak tests may be performed. Moisture penetration can be followed, for instance, with an integrated on-chip dew-point sensor (Tcheng et al., 2000). 'Accelerated' means that the packaged device is put into an autoclave filled with pressurized steam at high temperature (usually 130 °C, 2.7 atm, and 100% relative humidity) to accelerate the test (Chiao and Lin, 2002).

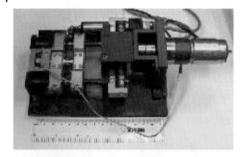

**Fig. 9.11** Example of an instrument used to test the tensile strength of a PDMS sample (top); PDMS samples (bottom) (Hanson, 2000).

### 9.2.4.3 **Other Tests**

Other tests include electrical integrity, performed by measuring the operating voltages while the device works. Optical device and interconnection tests involve for example the question of how much power is lost in relation to the source power. Tensile, shear, and torsional strength tests can be done on samples made of the same material as the package (Fig. 9.11). Ultrasound analysis using can be done to find defects in the structure.

Finally, reliability tests have to be used for lifetime prediction, failure testing, and physics-of-failure tests. For these tests, the packaged device can be subjected to elevated temperature for accelerating its degradation, to temperature cycling and temperature shock conditions to induce internal thermal stress, to vibrational stress to investigate ifs mechanical stability, and/or to salt atmosphere and hazardous chemicals for investigating corrosion.

### 9.3
### Influencing Factors in Packaging Design

The factors that influence the design of an element or perhaps the entire system can be defined as design factors. Several design factors usually have to be considered in any given design situation. Sometimes one or even several of them turn out to be critical, and when these are satisfied, the other factors need to be considered again. For example, the following list of design factors must often be considered in packaging:

- *Material requirements:* strength, fracture behavior, Young's modulus, corrosion, weight, wear, flexibility, stiffness, biocompatibility, bioresistance, chemical resistance, etc.
- *Thermal requirements:* thermal conductivity, thermal match with other materials, heating/cooling issues, etc.
- *Flatness requirements* (often in connection with the optical properties): average roughness (surface finishing), sides polished.
- *Interconnection requirements:* separation of electrical, optical, and fluidic interconnects, sealing properties, etc.
- *Optical requirements:* transparency at certain wavelengths, index of refraction, reflectivity, etc.
- *Electrical and magnetic requirements:* conductor or insulator, dielectric constant, magnetic properties, etc.
- *Process compatibility:* chemical compatibility, ease of metallization, machinability, etc.
- *Space requirements:* shape, size, volume, dead volume, etc.
- *Economic requirements:* reliability, utility, cost, safety, styling, maintenance, etc.

Some of these factors deal directly with the dimensions, the material, the processing, and the joining of the system elements. Other factors affect the configuration of the total system.

The size of the package is dictated by the need to interface to a larger structure, by limitations of traditional machining methods, or by the need for manual handling of the structure. After the package design is decided upon, the best manufacturing technique for the micromachined part inside has to be chosen. This sequence has to be followed because packaging costs contribute highly to the overall cost of any micromachined product and often do not respect size specifications. The many serial work processes involved justify the expensive packaging. Also, each device may require individual attention. Micromachining technique such as fusion bonding, anodic bonding, and other integrated encapsulation techniques, commonly used in bulk and surface micromachining, may transform more of the serial packaging steps into parallel batch.

## 9.4
## Factors Influencing Package Reliability

### 9.4.1
### Residual Stress

A mismatch between the mechanical properties of the materials creates a critical problem for the reliability of electronic packaging. For example, encapsulation of the device during packaging results in large residual stress, which is related to severe reliability problems; it creates problems in terms of cracked dies, passivation, and excess wear.

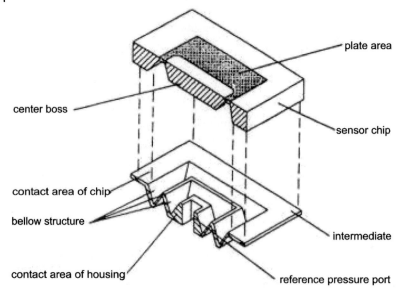

**Fig. 9.12** Device support for stress isolation (Madou).

Another problem is encountered when the microsystem has to operate in thermally changing environments. Here, the package has to be designed for a range of internal stress levels because internal stress is often a function of the temperature.

### 9.4.2
### Mechanical Protection and Stress Relief Structures

Rigidity, weight and size, and the possibility for repair or replacement by redundancy are the main considerations in mechanical packaging. The major causes of instability in packaging a microsensor or a microdevice are residual stresses and stray forces. Stress relief and stress isolation can be applied by placing a material or structure that can absorb the stress between the device and the package. Package stress is generated at two interfaces: between the sensor and the support and between the support and the package. A soft spring-like structure that deforms under the static load of the package can be used as a stress relief structure.

The approaches for stress relief at the sensor-support interface are:

– Use of materials with same thermal expansion coefficient.
– Use of low stiffness materials for interface bonding layers.
– Use of a symmetric layout of the structures to balance the effects of internal forces.

A micromachined support structure with V or U grooves (Fig. 9.12), can relieve up to 99% of the package stress.

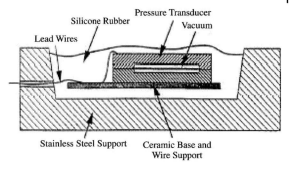

**Fig. 9.13** A catheter-tip piezoresistive pressure sensor embedded in silicone rubber and housed in a stainless-steel or titanium boat (Ko, 1995).

A stiff shell around the sensor embedded in a very pliant material can be used as a stress isolation structure. The microdevice can be isolated from stress if it is suspended by a complaint structure or inserted in a viscoelastic material and then housed in a stiff housing (Fig. 9.13).

Silicone rubber has superior adhesion to epoxy and a low elastic modulus, and therefore induces very little stress. However, it does not adhere to polished glass surfaces, sometimes swells in aqueous solutions, and has limited reparability. The use of thicker materials results in less residual stress; however, this may cause other problems, such as thermal stress or a too-bulky system.

### 9.4.3
### Electrical Protection and Passivation

The main considerations for the electrical protection and passivation of microdevices are electrostatic and electromagnetic shielding and moisture penetration. Moisture penetration is the most common failure mode for microdevices that must function in liquids or highly humid environments.

In humid environments, especially when the temperature fluctuates, moisture can penetrate into a microsystem. The absorbed moisture can condense in the cavity, causing current leakage and corrosion, generating noise, and leading to malfunction or even catastrophic failure. One solution would be to use hermetic packages made of impermeable glass, ceramic, or metal to ensure a barrier against moisture and ions for years. The permeability of metal, glass, and ceramics is very low, and these materials are considered impermeable to vapor at a thickness of 10 μm. Polymeric materials are not suitable for hermetic sealing (Fig. 9.10). Hermetic packages are expensive and difficult to manufacture.

In selecting packaging materials for moisture protection, one must also check the adhesion of the packaging material to the substrate, which is equally important. The interface between the device and the packaging material should have good adhesive bonding to prevent the creation of voids where vapor could condense, leading to electrolytic current and causing partial lift-off of the packaging material by corrosion. To obtain good adhesive bonds, the surface should be cleaned by the conventional cleaning methods that are used for integrated circuits, plasma, and ion beam etching. Interface stress and corrosion is also important in bonding; therefore, any galvanic couple formed by dissimilar materials is undesirable. Hermeticity is usually acceptable when the equivalent leak rate is below $5 \times 10^{-8}$ atm cc s$^{-1}$ air, as measured with a helium detector.

## 9.4.4
### Alignment During Bonding

An extra challenge comes from the requirement for good alignment between the device wafer and the support substrate (for instance, a glass or another silicon wafer) during bonding. One technique used for bonding alignment is to create holes in both the device wafer and the substrate, followed by placing them in a designed fixture to perform the bonding; however, the accuracy achievable by this method is only about 50 μm. Higher alignment accuracy ($\sim$2.5 μm straight) can be achieved with a bonding machine equipped with an in-situ optical alignment setup. Bower et al. (1991) worked on aligned wafer bonding using an infrared aligner modified to hold two imprinted wafers face to face while projecting an infrared image of the surfaces onto a viewing screen (Fig. 9.14).

The silicon wafers are etched, creating an array of V grooves, which are used to precisely align the wafers. After alignment, the wafers are brought in contact for bonding. Shoaf et al. (1994) developed an alignment technique for use in precise eutectic bonding of silicon structures. In this technique a silicon wafer is anisotropically etched to create V groves around the periphery of the structure to be bonded. Optical fibers are placed in the V grooves (Fig. 9.15). The fibers are placed orthogonally and are used as precision location keys for alignment. They are removed after bonding. The maximum misalignment is 5 μm for a 1×1-cm die. This technique allows precise bonding to an electronics die without using a microscope or micropositioners.

## 9.4.5
### Thermal Performance

Especially when fabricating many sensors, the temperature must be kept as low as possible. The thermal performance of the microsystem depends critically on the thermal properties of the many different materials of which the system is composed (Tab. 9.1). High thermal conductivity materials, special substrates with 'thermal vias', and thermal grease to reduce thermal contact resistance can be used as conventional heat removal techniques.

3D packages are almost always more difficult to manage thermally than 2D packages. Unlike 2D approaches, which can rely on sufficient package surface for cooling, 3D packages do not have this option. Heat is mostly removed by conductance, and improvements can be expected from the right material choice.

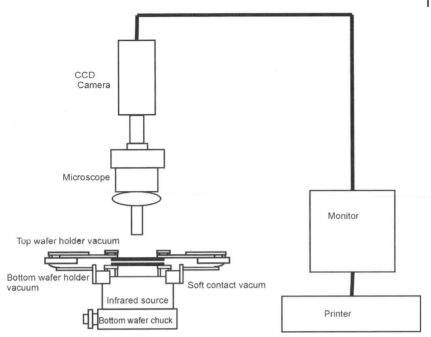

**Fig. 9.14** An infrared aligner system used for wafer bonding (adapted from Bower et al. (1991)).

**Fig. 9.15** Schematic representation of the anisotropically etched V-groove/optical-fiber alignment technique. Optical fibers are placed into the V grooves of the bottom silicon die. A second die with etched V grooves is placed on top of the first die (Shoaf et al., 1994).

The efficiency of convection cooling is directly proportional to the available surface area. Micromachining can be used to increase the surface area for convection cooling by machining, for instance, small channels in the substrates. For a micropump in a fluid environment, additional cooling may be obtained from the fluid itself as it comes in contact with the pump elements as it is forced through the

**Tab. 9.1** Thermal properties of common micromachining materials

| Material | Thermal conductivity (W m⁻¹ K⁻¹) | Specific heat (J g⁻¹ K⁻¹) | Thermal expansion coefficient (10⁻⁵ °C⁻¹) |
|---|---|---|---|
| Single-crystal silicon | 125 | 0.7 | 3.5 |
| Polysilicon | | | |
|   small grain size | 30–35 | | |
|   large grain size | 60–105 | | |
| Silicon dioxide | 1.5 | 0.8 | 0.5 |
| Silicon nitride | | | 4–7 |
| Polyimides | 0.15–0.25 | | 50 |
| Aluminum | 205 | 0.9 | 28 |
| Gold | 297 | 0.1 | 14.2 |
| Gold–silicon eutectic | | | 130 |
| Nickel | 92 | 0.4 | 13–15 |

system. A thermally efficient package needs to be designed by using simulation tools that can cope with board layout, the ambient temperature, and heat conductance.

### 9.4.6
### Chemical Resistance

Often, packaging materials must be chemically stable for long periods, because the microsystems inside the package are usually expensive. Microsystems operated in aggressive liquids or other harsh chemical environments need chemical protection. For example, silicon-micromachined multielement electrode arrays developed as electrochemical sensors are routinely exposed to harsh environments. The same is true for mass flow controllers, miniature flow injection analyzers, and other miniature fluid devices. The major cause of chemically related failure is moisture and ion penetration, which can result in corrosion, electrolytic conduction, and device failure. For example, in harsh environments, the membrane of pressure sensors is often isolated by a silicone oil-filled cavity that is capped or sealed with a second metal membrane. The metal cap transmits the pressure through the fluid to the pressure sensor.

Protection of the environment from the sensor operation is important in the biomedical as well as the food and drug industries. The main considerations are:

– reaction of the sensor with the environment
– toxic products from its operation
– sterility

Packaging of biomedical sensors requires special considerations respecting the operation, the environment, and the performance of the sensor.

## 9.4.7
### Protection During Packaging

Air-borne cutting fluids and solvents can attack the insulation and plastics during packaging. Moreover, sawing wafers can generate particles that can cause defects on a die and thereby reduce the yield. Generally, microsensors are delicate and often cannot survive this procedure. One solution is to protect the sensor structures with a cover prior to cutting. In general, good media isolation is necessary to protect the die during wafer sawing, die attaching, wire bonding, and molding. Various methods are available, including vapor-deposited organics (e.g., polyimide), silicone gel coating over the die, and use of a plastic or ceramic cap.

Many microsystems require miniaturized packages to retain their advantage of being small. To achieve the desired level of miniaturization, one can decrease the die size or the volume of the packaging. However, decreasing the die size may cause problems with cutting and handling, even though there would be more dies per wafer.

## 9.5
### Interconnections

The package and the interface to the environment determine the size and cost of the total product and the nature of the microdevice inside. Several methods have been experimentally used to interconnect systems fluidically, electrically, and optically. For instance, Fig. 9.16 shows a micromachined fluidic system including fluidic and optical interconnections. In this design, several layers of PMMA were made by using a $CO_2$ laser to form channels and layer-to-layer interconnects, and the layers were bonded together by thermal bonding.

### 9.5.1
### Fluidic Interconnections

To prevent the loss of fluids or undesired functioning of the system, all fluidic interconnections must be sealed hermetically. One of the most important parts of microfluidic systems is the fluid interconnection to the environment. A potential interconnect scheme involves the use of external tubing, as shown conceptually in Fig. 9.17.

This kind of interconnections requires precise manual fabrication and is not readily suited to mass production. Also important are the fluid interconnections between the different devices, which can be made macroscopically with tubing or with micromachined channels. Ordinary tubing has the disadvantage of a large dead volume, thereby reducing some of the advantages of microsystems. Therefore, it is necessary to develop techniques that allow connections of different microfluidic devices on a very small scale. The obvious way of integrating microfluidic devices is by vertical integration. Here, the different components are mounted on top of each other, as in Fig. 9.16, which shows a microfluidic system of five layers of PMMA bonded together. A great advantage of this approach is its simplicity.

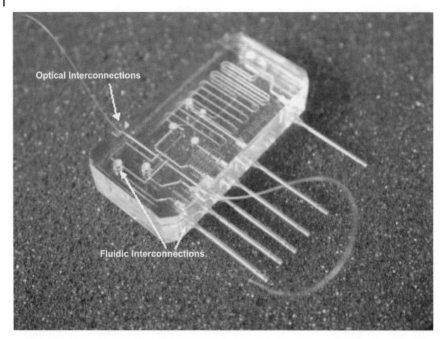

**Fig. 9.16** Microfluidic system including fluidic and optical interconnections (Detlef Snakenberg; MIC Denmark).

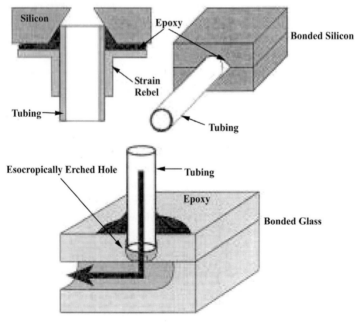

**Fig. 9.17** Conceptual off-chip fluidic interconnect schemes (Kovacs, 1998).

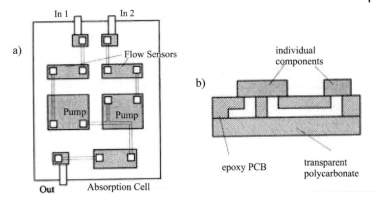

In 1    In 2

a)

Flow Sensors

individual
components

b)

Pump    Pump

Out    Absorption Cell

epoxy PCB    transparent
polycarbonate

**Fig. 9.18**  Mixed circuit board approach for microfluidic systems: a) top,
b) cross section (Koch et al., 2002).

Another technology is the so-called mixed circuit board, which consists of a top layer, a channel layer, and a bottom plate. The individual components are glued onto the bottom plate.

Another approach for interconnection systems was published by Gonzalez (Fig. 9.19). The assembly is based on stackable interlocking structures, similar to the concept of the popular Lego® toys. One type of interlocking structures are microfinger joints, made of two mating substrates that provide both breadboard interconnects and connectors to the outside world. The locking force is mainly determined by the friction of the two surfaces in contact, but the spring constant of the fins themselves contribute to holding the structure together. They are fabricated by using a diamond-tip circular saw. Grooves can also be fabricated by deep RIE (reactive ion etching) or anisotropic wet chemical etching.

Early approaches for making fluidic interconnections included insertion of tubing into etched or drilled fluid ports on the face or edge of the device and applying epoxy to form a seal (Fig. 9.20). A grommet or barbed fitting can be glued onto the device to provide mechanical stress relief. Some drawbacks of such an approach are the potential to clog the channels, the epoxy curing time, the irreversibility of the connection, the manual assembly steps, and the fact that epoxy is in contact with the working fluid. Deep reactive ion etching (section 6.5) has been used to couple a standard capillary tubing to silicon-based microfluidic devices.

Gray et al. (1999) proposed using deep reactive ion etching of bulk silicon to fabricate vertical interlocking structures, by which high density fluidic interconnections are provided between two microchannels containing substrates. The components and boards could be reversibly connected and disconnected (Fig. 9.21).

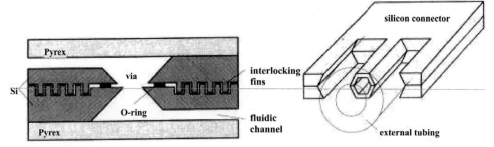

**Fig. 9.19** Cross section of a fluidic interconnect with a compression seal (left) and of a tube connector that allows the connection to the outside world (right) (Gonzalez et al., 1998).

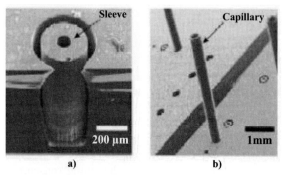

**Fig. 9.20** (a) Scanning electronmicrograph of a fluidic coupler with a 110 μm thick sleeve around the bore to prevent blocking of capillaries with adhesive. The silicon is cleaved to show the sleeve. (b) Scanning electronmicrograph of capillaries inserted into the sleeve couplers (Gray et al., 1999).

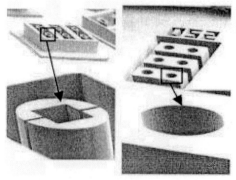

**Fig. 9.21** Scanning electronmicrograph of 250-μm diameter notched cylinders and holes. Each part has six fluidic connections (Gray et al., 2001).

9.5.2
**Electrical Interconnections**

Electrically isolated low-impedance feedthroughs between hermetically sealed regions or between devices are important for microsystems. The requirements include good insulation impedance, low feedthrough resistance, hermetically sealed small-size leads to go through layers of great thickness, and availability of multiple leads per cell while being fabricated on the wafer level by batch processes. The three main interconnection technologies are wirebonding (WB), tape automated bond (TAB), and flip chip (FC) technologies (Fig. 9.22).

WB uses wires to create connections, TAB uses tape leads, and FC uses bump drops of metal. WB is the cheapest technology and probably the most commonly used. TAB needs a smaller pad and pitch, has lower laboratory costs, and has better electrical performance. The disadvantages include the fact that the process is time consuming, and the design and fabrication of the special tape are expensive. FC ensures very small size, better electrical performance, improved thermal capabilities, and low cost. The disadvantages include the difficulty of testing bare dies and the limited availability of bumped chips. In addition, repair is difficult or impossible. Some of the feedthrough techniques developed in microsystem research to provide electrical connections from a hermetically sealed cavity to the outside are described below.

- *Thermomigration of aluminum column:* At sufficiently high temperatures, aluminum forms a molten alloy with silicon, and the aluminum in the alloy migrates along the applied temperature gradient.
- *P-n junction feed through:* The electrical connections from a sealed cavity of a silicon sensor to external terminals on a Pyrex glass substrate are p-n junction links that are alloyed to the metal connections at both ends.
- *Buried electrode feedthrough:* This technique is similar to the multiplayer metallization used in very large scale integration (VLSI); the electrodes are covered by the insulating layer through the wall of the sealed cavity.
- *Back contact:* The back of the wafer is etched to access p+ drain and source; the metallization can be made to connect the drain and source to the back of the wafer.

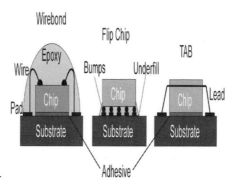

**Fig. 9.22**   Three interconnection technologies.

- *Sealed feedthrough channels:* Connecting electrodes are inserted into narrow channels, and then the channels can be sealed by glass frits or by sputtered insulators that can maintain the hermetic seal. This technique requires delicate handling and is expensive.
- *Connection through holes ('vias'):* Mechanical holes can be used to provide electrical connections between the two sides of the wafer or device. These holes must be small and placed at precisely defined locations to be compatible with the microcomponents or systems. They may be created by wet or dry etch, laser micromachining, ultrasonic drilling, electrochemical discharge drilling, or micromilling.

### 9.5.3
### Optical Interconnections

Micropackaging technologies for optoelectronic devices are used to reduce the assembly costs that are currently dominated by the fiber interfacing, which involves accurate positioning and fixing of an optical fiber in a critical alignment to the optoelectronic chip. The strong dependence of signal strength on the optical path length makes absorbance detection in small volumes difficult, because path lengths, and hence sensitivity, are generally reduced. The detection should be done perpendicular to the flow, and of course transparent materials are needed for optical detection. All these factors influence the optical interconnections. Hall et al. proposed the use of microetched silicon components for chip carriers with V grooves for passive fiber positioning (with an accuracy of $\sim 0.5$ µm) and solder bump technology for chip positioning and electrical interfacing (Fig. 9.23).

Another solution for optical interconnection could be the use of micromachined spring elements for fixing and alignment of the optical fiber (Fig. 9.24).

For facilitating optical alignment, Ceriotti et al. proposed a simple detection system with prealigned optical fibers. The chip made of PDMS is mounted on a stationary stage between the prealigned optical fibers, which are fixed by ferrules into a bulk piece that is attached to a positioning platform, so that the optical axis can be aligned with the stationary chip (Fig. 9.25).

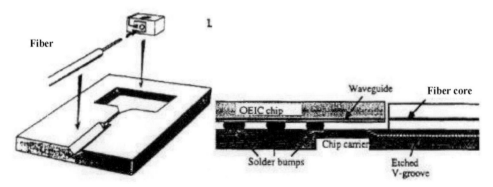

**Fig. 9.23** (left) Microfabricated silicon carrier for fiber optical assembly; (right) combined chip carrier/fiber carrier for passive alignment (Hall et al., 1994).

**Fiber-chip coupling**

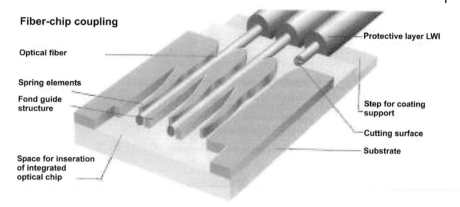

Fig. 9.24   Optical interconnection with springs (Gunstzsy of IMM, Mainz, Germany).

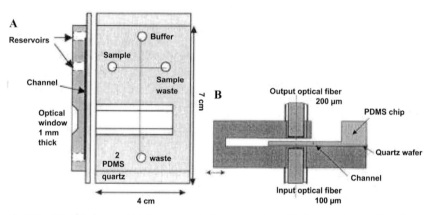

Fig. 9.25   (A) Chip layout: (right) top view, (left) cross section. (B) Cross section of chip alignment with respect to the optical axis (vertical dashed line) (Ceriotti et al., 2001).

Finally Boyle and Moore (2002) proposed using the property of deflection of micromechanical beams for passive alignment and fixing of optical fibers (Fig. 9.26).

## 9.6
## Comparison of Important Micromachining Materials

For the fabrication of an efficient micro device it is very important to know the mechanical properties of the material we are going to use. A rough comparison of important materials used in microtechnology is presented in Tab. 9.2.

As the table indicates, ceramic and glass are difficult to machine, and plastics are not readily amenable to metallization (see also section 6.3). Silicon has the highest material cost per unit area, but this cost often can be offset by the small feature sizes

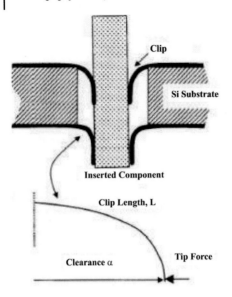

**Fig. 9.26** Cross section of the package structure (Boyle and Moore, 2002).

**Tab. 9.2** Comparison of materials commonly used in microtechnology.

| Material | | Cost | Frac-ture | Metal-lization | Machin-ability (common methods) | Dielectric constant | Young's Modulus E(GPa) | Thermal Conduc-tivity (W mK$^{-1}$) |
|---|---|---|---|---|---|---|---|---|
| Single crystal | Si | $$$$ | b,s | Good | Very good | 11.8 | 165 | 150 |
| | Quartz | $$$$ | b,s | Good | Poor | 4.4 | 87 | 7 |
| | GaAs | $$$$$ | b,f | Good | Poor | 13.1 | 119 | 50 |
| | Sapphire | $$$$$ | b,s | Good | Poor | 9.4 | 490 | 40 |
| Amor-phous | Fused silica | $$$– $$$$$ | b,f | Good | Poor | 3.9 | 72 | 1.4 |
| | Plastic | $$ | T,s | Poor | Good | – | – | – |
| | Paper/ cardboard | $$ | T,s | Poor | Fair | – | – | – |
| | Glass | $$–$$$$ | b,f | Good | Poor | 4.6 | 64 | 1.1 |
| Polycrys-talline | Alumina | $–$$$$ | b,s | Fair | Poor | 9.4 | 400 | –30 |
| | Alumi-num | $$$ | t,s | Good | Very good | – | 77 | –240 |

Note: b=brittle, t= tough, s=strong, f=fragile, $=very cheap, $$$$=very expensive.

**Tab. 9.3** Properties of some polymeric materials (Becker and Locascio, 2002).

| | PMMA | PC, high viscosity | PETG[a] | PE[b] | Polyimide | Styrene copolymer | Silicone |
|---|---|---|---|---|---|---|---|
| Melt flow (g/10 min) | 1.4–2.7 | 3–10 | | 0.25–0.27 | 4.5–7.5 | 1.4 | |
| Melt T (°C) | 85–105 | 150 | 81 | 98–115 | 388 | 100–200 | |
| Mold (linear) shrinkage | 0.001–0.004 | 0.005–0.007 | 0.002–0.005 | 0.015–0.05 | 0.0083 | 0.003–0.005 | 0.0–0.006 |
| Process T, (°C) injection | 163–260 | 294 | 249 | 149–232 | 390–393 | 182–288 | |
| Molding P, $10^7$ (Pa) | 3.4–13.8 | 6.9–13.8 | 0.69–13.8 | 3.5–10 | 2–13.8 | 3.5–13.8 | |
| Hardness, Rockwell | M68–105 | M70–75 | R106 | | E 53–99, R129, M95 | M80, R83 | |
| Coef. linear thermal expansion, $10^{-6}$ (°C) | 50–90 | 68 | | 100–220 | 45–56 | 65–68 | 10–19 |
| Thermal conductivity; $10^{-4}$ g cal-cm s$^{-1}$ cm$^{-2}$ °C | 4–6 | 4.7 | | 8 | 2.3–4.2 | 3 | 3.5–7.5 |
| Dielectric strength, 0.003175 m specimen MV m$^{-1}$ | 16–20 | 15–16 | | 18–39 | 16–22 | 17 | 16–22 |

a) polyethylene terephthalate glycol
b) polyethylene, branched homopolymer

**Tab. 9.4** Properties of heat sink materials at room temperature.

| Material | Thermal conductivity $[W\ m^{-1}\ °C^{-1}]$ | Density $[kg\ m^{-3}]$ |
|---|---|---|
| Aluminum | 220 | 2.7 |
| SiC/aluminum | 170 | 3 |
| Boron/aluminum | 145 | 2.7 |
| Copper | 400 | 8.96 |
| Cu-coated graphite/Cu | >400 | 5.3 |
| Cu–Mo–Cu | 170–210 | 10.08 |
| Cu–Invar–Cu | 164 | 8.4 |

possible with silicon. Silicon is especially preferred in thin-film applications. If the substrate merely functions as a support, glass, ceramic, or even plastics and cardboard become options. If the substrate has a mechanical function, Si is an excellent candidate. If the substrate must have good optical properties, materials such as GaAs and poly(methylmethacrylate) (PMMA) are candidates.

The packaging requirements are more or less stringent, depending on the type of device that has to be packaged. For a sensor in aqueous solutions for example, a ceramic substrate requires no protection of its sides; a Si sensor, on the other hand, is difficult to insulate and to package, because a conductive medium might short the electrical sensor signal via the conductive Si sidewalls. The operating temperature of the system influences the choice of a material having high conductivity or not. Ceramic packaging, with its high thermal dissipation, could be used to address this problem. Unfortunately, ceramic packages are much more expensive than polymer laminate packages. Moreover, the dielectric constant of ceramic materials is anywhere from 1.5 to 10 times higher than those of most polymeric materials, and the dielectric constant is directly proportional to the incidence of cross-talk between signals and losses of signal intensity. Finally, ceramics are brittle; they are not very resistant to torsional and impact stresses, and the forces acting on them have to be precisely directionally oriented to prevent breakage.

It is evident that polymers present a more attractive alternative to ceramics from the viewpoint of cost, size, weight, performance, and mechanical performance. Tab. 9.3 presents the principal properties of the most common polymeric materials.

For moderately high thermal operating environments, the use of some cooling mechanism can be necessary for meeting thermal dissipation requirements in polymer packaging. The cooling systems can include conduction, convection, and radiation cooling. One of the best ways to provide adequate thermal dissipation is heat sinking (Gallager et al., 1998). Tab. 9.4 lists some of the materials employed as heat sinks and their properties.

The problem with these package configurations is that they are expensive, heavy, and large. Additionally, the thermally resistant interfaces between the polymer and the heat sink diminish the effectiveness of thermal dissipation. However, using a metal substrate that is electrically insulated by a polymer dielectric that has been highly filled with a thermally conductive ceramic powder, it is possible to better integrate the heat sink directly to the package. Thus it is possible to minimize thermally resistive interfaces. Finally, the choice of material depends on its chemical resistance properties. Generally, some plastics are good chemically resistant materials, but silicon is not chemically resistant and has to be protected from harsh environments.

## 9.7
## References

PH. ARQUINT et al., *Flexible polysiloxane interconnection between two substrates for microsystem assembly*, 8th Intl. Conf. Solid-State Sensors and Actuators, Stockholm, Sweden, **1995**, 263–264.

H. BECKER, L. E. LOCASCIO, Polymer microfluidic devices, *Talanta* **2002**, *56*, 267–287.

R.W. BOWER, M.S. ISMAL and S.N. FARRENS, Alligned wafer bonding: a key to three dimensional microstructures, *J. Electron. Wafer* **1991**, *20*, 383–387.

P. BOYLE and D.F. MOORE, Micropackaging using thin films as mechanical components, **2002**, IEEE, San Diego, CA, USA.

L. CERIOTTI et al., Visible UV detection through square deep PDMS channels, *Micro Total Analysis Systems* **2001**, 339–340.

J.-Y. CHEN, L.-S. HUANG, C.-H. CHU, C. PEIZEN, A new transferred ultra-thin silicon micropackaging, *J. Micromech. Microeng.* **2002**, *12*, 406–409.

M. CHIAO, L. LIN, Accelerated hermeticity testing of a glass-silicon package formed by rapid thermal processing aluminum-to-silicon nitride bonding, *Sensors and Actuators A* **2002**, *97/98*, 405–409.

C. GALLAGER et al., Materials selection issues for high operating temperature (HOT) electronic packaging **1998**, IEEE, Proceedings of the High-Temperature Electronic Materials, Devices and Sensors Conference, 22–27 Feb 1998, IEEE, 180–189.

C. GONZALEZ, S.D. COLLINS, R.L. SMITH, Fluidic interconnects for modular assembly of chemical microsystems, *Sensors and Actuators B* **1998**, *49*, 40–45.

B.L. GRAY, S.D. COLLINS, R.L. SMITH, Novel interconnection technologies for integrated microfluidic systems, *Sensor and Actuators* **1999**, *77*, 57–65.

B.L. GRAY, S.D. COLLINS, R.L. SMITH, Interlocking mechanical and microfluidic interconnections fabricated by deep reactive ion etching, *Micro Total Analysis Systems* **2001**, 153–154.

J.P. HALL, M.Q. KEARLEY et al., Passive fibre interfacing to optoelectronic devices using hybrid micropackaging, Institution of Electrical Engineers, London, UK, Seminar Digest **1994**/043, P9/1–6.

D.E. HANSON, A mesoscale strength model for silica filled polydimethylsiloxane based on atomistic forces obtained from molecular dynamics simulations, *J. Chem. Phys.* **2000**, *113*, 7656.

W.H. KO, Packaging of microfabricated devices and systems, *Materials Chemistry and Physics* **1995**, *42*, 169–175.

M. KOCH, A. EVANS, A. BRUNNSCHWEILER, *Microfluidic Technology and Applications*, Research Studies Press (RSP), Baldock, Hertfordshire, UK, **2000**.

G.T.A. KOVACS, *Micromachined Transducers Sourcebook*, WCB McGraw-Hill, New York, **1998**.

U.D. LARSEN, *Microliquid Handling – Passive Microfluidics*, Ph.D. Thesis, July **2000**, Mikroelectronic Centret, Lyngby, Denmark.

M. MADOU, *Fundamentals of Microfabrication*, Ron Powers, Boca Raton, Florida, USA, **1997**.

W.K. SCHOMBURG et al., Assembly for micromechanics and LIGA, *J. Micromech. Microeng.* **1995**, *5*, 57–63.

S.E. SHOAF et al., Aligned Au-Si eutectic bonding of silicon structures, *J. Vac. Sci. Technol.* **1994**, *A12*, 19–22.

K. M. STRINY, Assembly techniques and packaging of VLSI devices, in *VLSI Technology*, ed. S. M. SZE, McGraw-Hill, New York, **1988**.

Y. TCHENG, L. LIN, K. NAJAFI, Fabrication and hermeticity testing of a glass-silicon package formed using localized aluminum/silicon-to-glass bonding, **2000**, IEEE.

# 10
# Analytical Chemistry on Microsystems

Jörg P. Kutter and Oliver Geschke

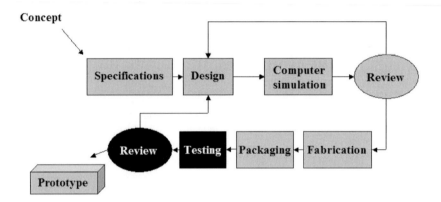

One of the main applications of microfabricated chemical systems so far is analytical chemistry. This branch of chemistry is concerned with gathering qualitative and quantitative information on all kinds of samples, often with only a small amount of sample to work with or with the analytes of interest present in very low concentrations [1]. At the same time, the matrices (i.e., the 'stuff' containing the analytes of interest) can be very complex and challenging, for example, wastewater or body fluids. Analytical chemistry typically asks the following questions, and, of course, tries to find answers to them:

- What is in the sample? Identification of components (X, Y, Z, etc.).
- How much of substance X is in the sample? Quantification.
- In what form is X present? Speciation – often necessary in toxicological assays.
- How did X get into the sample? This question is very important for forensic analysis, environmental pollution analysis, and investigations on the metabolism of drugs and toxic substances.

To successfully answer all these questions, the analytical process consists of several steps, all of which have their own particular challenges. Typically, the following main steps are necessary:

*Microsystem Engineering of Lab-on-a-chip Devices*
O. Geschke, H. Klank, P. Telleman
Copyright © 2004 Wiley-VCH Verlag GmbH & Co. KGaA, Weinheim
ISBN: 3-527-30733-8

- sampling
- aliquoting/injecting/metering
- sample preparation (filtering, enrichment, reaction, labeling, etc.)
- separation
- detection
- sensing
- evaluation

Traditionally, between these steps much handling and transport of untreated and treated samples occurs, greatly increasing the risk of contamination and loss and therefore of less trustworthy analytical results. In the 1980s the idea was born to build systems incorporating all necessary parts for performing a chemical analysis. Such devices were called total analysis systems (TAS). One of the main motivations for such devices was the possibility to take samples on site and analyze them in the field or at the patient's bed, instead of having to bring samples to a central lab. Thus, transportability was a major requirement. However, in reality, many of those devices were rather bulky and moving them required the help of heavy machinery. TAS was only a first step – the trend was clearly set to not only incorporate all functions necessary for a chemical analysis, but, at the same time, to miniaturize the equipment as much as possible. This led naturally to the concept of microTAS (µTAS) [2–4]. Later, these systems came to be called 'lab-on-a-chip' systems, to account for the possibility that not just analytical tasks could be performed on such a device, but also other chemical functions, such as synthesis. However, the scope of this book is restricted to analytical applications.

What, then, are the advantages of performing analytical chemistry on microchips? Some are obvious. Because of the small dimensions of the channels, µTAS devices can potentially work with very small amounts of sample. This is a tremendous advantage in, for example, medical diagnostics, where it is preferable for both the patient and the medical assistant to be able to work with 1–5 mL of blood instead of half a liter, or in drug discovery, where only very small amounts of new drug candidates are synthesized, to keep costs down. In addition, the use of expensive reagents is drastically reduced in microsystems, and less waste is produced. The same is true for energy consumption and (unwanted) heat generation. Also, it should be possible to perform many analyses much faster on microdevices, because of the reduced diffusion distances. Although not much can be done to accelerate the kinetics of a given chemical reaction, all diffusion-limited processes are faster than in conventional systems, in which diffusion distances are considerably longer. Another possibility of speeding things up in microsystems is parallelization, for which they are eminently suited. Machining a single channel takes basically the same amount of work as machining an array of channels to allow reactions, separations, etc. to take place in parallel. Together with increased speed because of reduced diffusion distances, parallelization allows microdevices to process a large number of samples, orders of magnitudes more than conventional systems are capable of. This is extremely important in drug discovery, because many possible drug candidates have to be tested against many possible targets. Further advantages of miniaturized analytical systems are being discovered as our understanding of how liquids perform on a small scale improves. As shown in previous chapters, many microsystem implementations make use of such phenomena as laminar flow and the predominance of diffusion to perform analytical tasks in a way not accessible to conventional equipment. Finally, the production cost for individual microdevices can be decreased enough that one-time use is not only reasonable from an analytical point of view, but also economically viable. In particular, polymer materials are increasingly being used to produce such devices for, e.g., diagnostic purposes, where one-time use is mandatory to avoid contamination and the necessity to elaborately clean the microchip for a second use.

The list of advantages is rather impressive, but what are the disadvantages? What does one have to keep in mind when designing microsystems for analytical purposes? We must not forget that the volumes encountered in microsystems are very small, and, depending on the concentration of the solution, only very few molecules might be in the volume we have to work with (Fig. 10.1). Let us assume we have a cube with side length $d$, representing our sample container. If we further assume that the sample concentration is 1 nM (this is low, but often encountered in medical and environmental trace analysis), then Fig. 10.1 gives us the volumes and the number of molecules in this volume for different values of $d$. A cube with a side length of 1000 μm (1 mm) represents 1 μL of volume and about $10^9$ molecules at the given concentration. For a cube of 1 μm side length, the volume is only 1 fL ($10^{-15}$ L) with only about one(!) molecule within this volume. Clearly, it is impossible to 'work' with one single molecule, because it is only present in a statistical sense. Even if we were sure to find this molecule in our probe volume, handling losses (e.g., by adsorption) would prohibit successful analysis, not to speak of the challenges of detecting a single molecule. From these facts we have to conclude that it is not reasonable to work in the 1 μm regime, but instead we need to use larger volumes or to increase the concentration of the sample prior to further analysis. Fig. 10.1 also shows the diffusion time of a typical molecular species as a function of distance. A typical small molecule can diffuse a distance of 1 μm in less than 1 ms, however it takes the same molecule

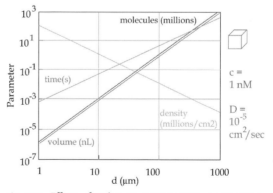

**Fig. 10.1** Effects of scaling on some important parameters (adapted from [5]).

more than 8 minutes (500 s) to diffuse a distance of 1000 μm. Obviously, from this point of view it is preferable to work at small scales, since diffusion is often the only transport mechanism available, and rapid analyses are the goal. Thus, it becomes apparent that working in a size range between 10 and 100 μm is a good compromise, which is exactly why many microsystems for analytical purposes are within this range. This is not to say that there are no reasons to explore the extreme ranges some more. In fact, there are some exciting applications on chips that have channels that are extremely shallow (smaller than 1 μm) or much larger than 100 μm (up to 1 mm). Research in these areas can be rewarding as long as one keeps the limitations (with respect to concentration and diffusion time) in mind.

In order to create μTAS devices a large number of functional elements need to be available, in essence, a toolbox providing all the functionality available in traditional benchtop systems. A nonexhaustive list of such functional elements could include channels, fluidic connections, pumps, valves, dosing units, reactors, mixers, physical filters, sorters, coolers, heaters, physical and chemical sensors, separation and extraction units, light sources, detectors, electronics, and chip-to-world interfaces.

Many of these elements have been described in previous chapters. In this chapter, we take a closer look at some of the more complex elements and, especially, at systems consisting of more of these elements, in a modular or in an integrated fashion. Among the many approaches to tackling an analytical problem, we focus on three very common strategies: a) for direct measurements of one or a few components with no or little sample preparation, chemical or biological *sensors* are used; b) for measurements of one or a few components for which some treatment of the sample is required (e.g., cleaning, derivatization, reaction) *flow injection analysis (FIA)* coupled to a suitable detection system is often used; c) for more complex samples it becomes necessary to *separate* the components into individual bands by means of techniques such as electrophoresis and chromatography, also coupled to a suitable detection system.

## 10.1
### Sensors and Sensor Systems

As discussed in section 4.4, sensor systems can perform chemical analyses of samples. Since most of the chemical analyses done today are in aqueous solutions, we focus on those. We should mention, however, that many different gas sensors are used in everyday life, just one example being the lambda sensor ($O_2$ sensor) used to optimize fuel combustion in automobiles. Other types of oxygen sensors are used for wastewater monitoring; however, their construction is far different from that of the lambda sensor.

In 1956 L. C. Clark [73] proposed an electrochemical oxygen sensor based on an amperometric electrode, covered with a gas-permeable membrane. He set the working electrode to a potential at which oxygen is reduced to hydroxide.

$$O_2 + 2\,H_2O + 4\,e^- \rightleftharpoons 4\,OH^-$$ (10.1)

**Fig. 10.2** µTAS-system for water analysis, measuring phenolic compounds, dissolved oxygen, nitrate, pH value, and amber color. Left: components, right: mounted system. Taken from [74].

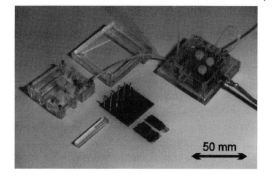

50 mm

Hydrogel (Cl⁻)          Silicone-membrane

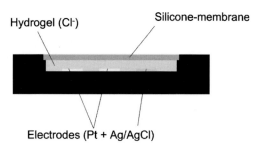

**Fig. 10.3** Cross section through an oxygen sensor.

Electrodes (Pt + Ag/AgCl)

Because all gases permeable through the membrane are electroinactive at that potential or only present in trace amounts, the overall signal depends only on the oxygen content. However, certain samples, e.g., those with very high $H_2S$-concentrations, are problematic: because the $H_2S$ is reduced as well and thereby adds to the signal, the oxygen concentration is actually lower than that measured with a Clark-type sensor. As stated above, miniaturization of such sensors is difficult because of the miniaturization of the reference electrode. However, approaches have been made to integrate a miniaturized oxygen sensor into a miniaturized flow system (Fig. 10.2).

The oxygen sensor itself consists of an array of parallel switched microelectrodes to reduce the 'consumption' of oxygen, because transport through the membrane is the slowest step of the overall reaction. If only a small fraction of the diffused oxygen is consumed at the electrode, the concentration underneath the membrane is almost constant and almost the same as on the other side of the membrane in the sample (Fig. 10.3).

An electrolyte is needed to maintain a constant chloride ion concentration, and it is useful to use a gel rather than a liquid as the electrolyte. This gives much better mechanical support for the oxygen-permeable silicone membrane. Although the actual electrodes can be fabricated by conventional cleanroom technology, the electrolyte and the membrane have to be applied manually.

In addition to those amperometric sensors, potentiometric sensors play an increasing role in biochemical microsystems. As mentioned in section 4.5, field ef-

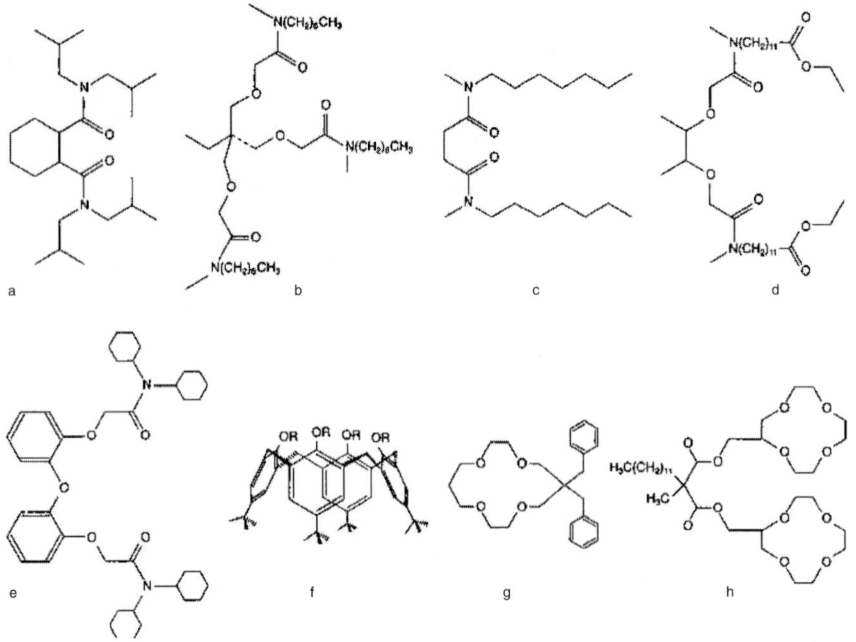

**Fig. 10.4** Some commonly used ionophores and their specificities: a) [ETH1810] for Li$^+$; b) [ETH227] for Na$^+$; c) [ETH1117] for Mg$^{2+}$; d) [ETH1001] for Ca$^{2+}$; e) [V163] for Ba$^{2+}$; calix[4]arene for alkali ions; g) 6,6-dibenzyl-14-crown-4 for Li$^+$; and h) bis[(12-crown-4) methyl]dodecylmethylmalonate for Na$^+$ [75].

fect transistors can be made sensitive not only to a shift in pH but also to a variety of different ions, leading to their name of ion-selective field effect transistors, ISFET. Most often, the FET is covered with so-called liquid membranes. These can be based on various hydrogels (solgels) or contain modified or unmodified polyvinylchloride (PVC). Generally, ion-selective PVC membranes consist of 66% softener, 33% PVC, and 1% ionophore, which is responsible for the sensitivity. Ionophores are often cage-like molecules that host the species of interest. The best known and probably best described ionophore today is valinomycin, a molecule with very high affinity to potassium. The potassium ions basically have the right charge and size to fit into the cage built up by the valinomycin. Similarly charged ions such as sodium or hydrogen are too small to get trapped inside the cage, which is also called a clathrate (Greek *clatherus*=cage). Other frequently used ionophores are shown in Fig. 10.4.

As indicated by the names of some of the ionophores, many were created artificially at universities such as ETH-Zurich ([ETH1810] or [ETH227]) or by drug companies to target specific molecules of interest such as calcium, magnesium, and fluorine.

## 10.2
## Biosensors

A biosensor is a sensor that includes a biological recognition element, most often immobilized on top of a chemical sensor. These recognition elements can vary in size from whole organisms or cells (both living and dead) down to enzymes or antibodies.

Even the canary once used by coal miners can be considered biosensors, because, as soon as it stopped singing, this meant that the ambient air contained either too little oxygen or, even worse, toxic gases. This might sound old-fashioned; however, in state-of-the-art wastewater treatment plants living fish are often used to monitor the water quality – if they die, the water quality is certainly not good enough for humans to drink. Also, the swimming behavior of daphnia, a small crab species, has been used for monitoring water quality [76].

The unique recognition elements formed in nature are the key to extremely specific, sensitive sensors. For example, an increasing number of people suffer from allergies – their immune system is extremely sensitive to certain materials, e.g., nickel. For analytical purposes, one could imagine these people dipping a finger into a solution containing trace amounts of nickel. Depending on the concentration, they will develop a larger or smaller area of itchy skin. This is of course just a picture to illustrate the function of a biosensor and fortunately, we do not need to take this approach – antibodies and enzymes are now commercially available and can be used in the fabrication of biosensors.

We describe some systems based on the enzyme glucose oxidase here, because this common enzyme offers a variety of detection possibilities.

Because enzymes are not reagents, but highly specific catalysts, they are not used up during a reaction. The activity of a specific enzyme is very much dependent on the enzyme source and its microenvironment. Its activity can vary from very high in one solution to zero (due to its complete denaturation) in another. Because the 'same' enzyme from different sources can differ significantly in weight, and often only a fraction of the protein is active, it does not make sense to describe a concentration in milligrams per milliliter. The company Sigma alone offers 10 different types of glucose oxidase in their product catalogue. To compare different products, the term *activity* of an enzyme, given in units (*u*), is used. One unit of enzyme activity is described as the rate of its reaction under certain conditions such as pH, temperature, and buffer concentration.

This means also that enzymes other than glucose oxidase have other specifications, meaning that a unit is defined at a different pH value, a different temperature, and a different buffer concentration. This information is most often provided by the manufacturer as specifications for a given lot of enzyme, because the activity of different lots of the same enzyme can differ. For example, a unit definition from Sigma is *"One unit will oxidize 1.0 μmole of β-D-glucose to D-gluconolactone and $H_2O_2$ per min at pH 5.1 at 35 °C, equivalent to an $O_2$ uptake of 22.4 μl per min."* If the reaction mixture is saturated with oxygen, the activity may increase by up to 100% (Sigma-Aldrich documentation).

Glucose oxidase catalyses the oxidation of glucose:

$$\text{glucose} + O_2 \quad \rightarrow \quad \text{gluconolactone} + H_2O_2 \tag{10.2}$$

To analyze a sample for glucose (e.g., a blood sample from a diabetic), more than one strategy can be used. For instance, one could measure the consumption of oxygen during the reaction with an oxygen sensor. However, because this entails a small decrease in a relatively large background signal, this measurement has a very bad signal-to-noise ratio, resulting in a relatively large measurement error. Basically, this measurement would be like weighing a large container ship twice, once with the captain on board and a second time without the captain, and then calculating the weight of the captain as the difference. Much more accurate is the measurement of the hydrogen peroxide produced. $H_2O_2$ can be monitored either directly, amperometrically, or by mixing the solution with alkaline luminol solution and using iron(III) ions as catalyst. This approach enables very sensitive measurement, because the reaction produces light in direct proportion to the hydrogen peroxide concentration which is itself directly proportional to the initial glucose concentration. Certainly, this measurement technique is applicable on the microscale, because a channel system, providing the different solutions, mixers, and optical waveguides, as well as photodiodes, can be relatively easily integrated into a silicon microsystem. However, one major downside of this promising technique is that the signal is produced after reagents are added, which means that waste is produced to obtain the information. Even if a microsystem consumes considerably less reagents than a conventional macrosystem, we still run the risk that excess waste is produced to ensure that a sample is 'clean'. This might not be crucial for the proposed glucose measurement with luminol, but certainly some reactions require much more hazardous reagents.

But let us go back to the amperometric determination of glucose. Because the reaction produces hydrogen peroxide proportional to the initial concentration of glucose, the hydrogen peroxide concentration itself can be monitored. This is most often done by an electrochemical oxidation of the substance with a platinum electrode. The platinum electrode has to be polarized to about 750 mV against Ag/AgCl/3M KCl to oxidize the $H_2O_2$ according to

$$H_2O_2 \quad \rightarrow \quad H_2O + O_2 + 2e^- \tag{10.3}$$

The measured current corresponds directly to the number of transferred electrons, which is proportional to the concentration of hydrogen peroxide and thus to the initial glucose concentration. In a way, this reaction is reagentless – no chemicals must be added to obtain a signal. But this method also has some drawbacks, one of them being the already mentioned unstable miniaturized reference electrode. Additionally, electrochemical interferences are a major concern of electrochemical measurements: substances such as paracetamol and ascorbic acid, which are often found in blood samples, are also oxidized at the measurement potential, leading to a higher background signal. If this occurs, the measured glucose level is too high. To circumvent such measurement errors, many devices consist of at least two almost identical sensors to enable differential measurement between an enzyme-coated sensor and a similar sensor without enzyme or with denatured enzyme.

One other promising way of getting rid of unwanted interferences is to use an electron mediator. Those mediators are reversible redox systems, such as ferrocene or ferro/ferricyanate, which can be oxidized/reduced at potential differences close to zero.

One very interesting issue concerning the miniaturization of biosensors is in attaching the recognition element to the chemical sensor. Attaching biomolecules to a microstructure such as an optical fiber or a microelectrode most often requires spatial resolution. Only very few methods can be applied to pattern the sensitive layer. Light-induced activation/polymerization, most often through masks, is one common way of achieving patterns. A second, rapidly evolving technique is pin-printing. This technique is often used to create arrays of tiny (nanoliters to picoliters) droplets to fabricate, for instance, DNA arrays. Pin-printers consist of one or several pins that are automatically dipped into a fluid by a robot and afterwards gently pressed against a substrate such as glass, ceramic, or polymer. Surface tension, capillary forces, and of course pin size determine the amount of liquid transferred from the pin to the substrate.

One very elegant method of creating spatially resolved coatings is electropolymerization. By dipping an electrode into an aqueous solution that contains an electropolymerizable monomer such as pyrrole, aniline or thiophene and the biological recognition element, the electrode can be coated by forcing a current through it. With pyrrole, the pyrrole becomes oxidized to form a radical cation that combines with an additional radical cation to form a dimer, releasing two protons. Recombination of the dimers and addition of further pyrrole molecules lead to polymer chains of up to 300 monomer units. Because the solution also contains the recognition elements, some of these become entrapped in the electropolymer. Because the polymer grows only on the electrode that is connected to the power source, all other electrodes remain uncoated. By repeating this procedure with other recognition elements and connecting different electrodes on the same substrate to the power source, multisensors can be created in a relatively short time. Already in 1994, H. Meyer et al. [77] fabricated a microelectrode array consisting of 400 individual addressable platinum microelectrodes having a surface area of 50 $\mu$m$\times$50 $\mu$m, arranged in an area of a square inch. These electrodes can be used in both potentiometric and amperometric mode.

## 10.3
## Flow Injection Analysis

Flow injection analysis (FIA) was developed in the mid-1970s and is an important precursor to the development of micrototal analysis systems. Besides helping pave the way for $\mu$TAS it is of course also a powerful analytical technique in its own

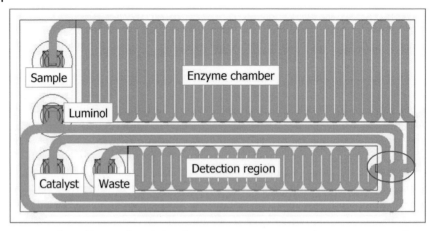

**Fig. 10.5** Channel layout of a FIA chip showing an enzyme reactor, a laminating mixer, and a detection area (courtesy of A. M. Jørgensen, MIC).

right, and an enormous number of analytical approaches now use FIA in its original or a modified form. For more details on FIA please see references [6, 7]. In its basic implementation FIA is an arrangement, in which a defined amount of sample is injected into a carrier stream that contains reagents with which the sample reacts. Simultaneously, the sample plug disperses in a controlled way as it moves downstream, where a time-dependant signal is registered by a detector. Because all aspects concerning timing in this system, i.e., flow rate, arrival time at the detector, reaction rate, dispersion rate, etc., are known or can be controlled reproducibly, it is not necessary to wait for steady state to be reached. Every part of the resulting time-dependent signal can be exploited for quantification, as long as reproducible timing is guaranteed. And as long as this main principle is adhered to, the FIA system can be expanded to add any number of extra functionalities. For example, if the chemistry dictates it, several different reaction steps can be implemented in sequence. The sample may also be submitted to cleaning or extraction/dialysis steps prior to reactions. Because a large number of analytical procedures using FIA principles have been implemented on microchips recently, we limit ourselves to a closer look at only two examples.

Determination of glucose in samples from bioreactors (fermenters) can be implemented as a flow injection analysis on microchips (Fig. 10.5). Although the first step should be dialysis to extract glucose from the broth in the fermenter tank, this has not yet been added to chip functionality. The next step then is conversion of glucose into hydrogen peroxide, which takes place in an enzyme reactor. The enzyme reactor can be designed as a long meandering channel segment, in which the enzyme is immobilized on the wall. After a plug of glucose is injected into the carrier buffer stream, it is converted to hydrogen peroxide as it passes through the enzyme reactor part. The flow emerging from the reactor is then laminated with two reagent flows (luminol and iron(III) ions) to initiate a re-

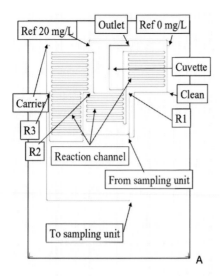

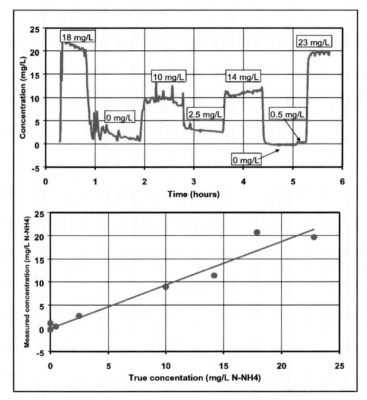

**Fig. 10.6** A) Layout of a FIA chip for detection of ammonium in wastewater. B) Measurement results for different ammonium concentrations and resulting calibration plot (courtesy of MicroChem consortium).

action that emits light. This light can be registered by suitable detectors, and the amount of light is proportional to the initial concentration of glucose. The first demonstration of this analytical procedure on microchips used off-chip injection devices and off-chip pumping to push the carrier and reagent streams through the microfluidic system.

Another example of µFIA is a chip for analysis of ammonium in wastewater [8, 9]. Here, sampling occurred as ammonium ions in the wastewater were extracted into a carrier stream via a dialysis membrane. The channel layout (Fig. 10.6 a) consists of a series of channels and intersections where different reagents are added to the carrier stream containing the sample. The dimensions of the various channel segments are chosen to allow fast mixing of the reagents by diffusion and sufficient time for reaction without the cost of too high counterpressures. Results from continuous-flow measurements at different concentrations of ammonium are shown in Fig. 10.6 b. The chip contained reference solutions of ammonium to allow on-chip calibration. Both reference and sample solutions formed a dye upon reaction, which was photometrically detected in an optical cuvette built into the chip. Challenges in this system were manifold, ranging from investigating the effect of channel geometry on dispersion, to stability of chemical solutions, and also to incompatibility between the chosen chemistry and the chosen chip materials.

FIA solutions are quire amenable to implementation on fluidic microchips. However, a few things need to be considered. Although miniaturization can reduce consumption of chemicals (samples, reagents) and help speed things up by decreasing the necessary diffusion distances, not much can be done to favorably change the kinetics of a chemical reaction. It is therefore not the time gain that is the main important issue when miniaturizing FIA, but the drastically reduced consumption of materials, which allows assays involving expensive chemicals or limited amounts of sample to be performed. A certain amount of time may be saved by the possibility of analyzing many samples in an automated way with µFIA, either in parallel or in series. The other thing to keep in mind when implementing microchips for FIA applications is to carefully design the channel network layout and the channel geometry so as to optimize system performance with respect to mixing time, flow rate, fluidic resistance, and back pressure and to minimize unwanted and unnecessary dispersion. These parameters need to be optimized to obtain good signals (good signal-to-noise ratios) and to ensure reproducible timing; otherwise, the FIA approach is not feasible and the interpretation of the data does not give analytically reasonable results.

## 10.4
## Separation Techniques

For complex samples, where many components are present, it is often necessary to separate the mixture into the individual components before they can be detected and/or quantified. The most commonly used separation techniques in traditional analytical systems as well as in analytical microdevices are electrophoresis and chromatography. Electrophoresis can be used for charged molecules (ions), which, in the simplest mode of electrophoresis, are separated in an electric field according to their charge-to-mass ratios. Besides this electrophoretic motion an additional movement is often imparted by the electroosmotic flow (see section 4.1). Chromatography, on the other hand, is suitable for separation of both charged and uncharged molecules. The separation makes use of physicochemical differences between the components, such as different affinities for a stationary phase or different distribution coefficients between a stationary phase and a mobile phase. Chromatographic techniques require the presence of a stationary phase through which the mobile phase and the mixture to be separated are percolated. Movement of the mobile phase is induced by pressure or electroosmosis. Both electrophoresis and chromatography have several modes and variants, making them excellent tools for a variety of separation problems.

To perform a successful, useful separation on a chip three steps need to be executed: a) injection, in which a defined portion of sample (an aliquot) is introduced into the separation system by pressure or electrokinetic means; b) separation, during which the different components in the sample are moved through the separation system and separated into individual bands or zones; c) detection, by which the individual bands are registered at the outlet of the separation system based on various physical principles. Even though the first analytical microdevices were chromatographic systems [10, 11], it was soon realized that it would be easier from an engineering point of view to create electrophoretic systems first. In principle, a microchip with a cross layout and electrodes in buffer reservoirs at the ends of the channel sections should allow electrokinetic injection and electrophoretic separation to be performed. Indeed, one of the first successful reports on separations on microchips described the electrophoretic separation of a mixture of fluorescent-labeled amino acids (Fig. 10.7). A small amount of sample (picoliters to nanoliters) was injected by the injection techniques described in section 4.2 and was transported down a main channel, where separation occurred because of the different migration velocities of the components in the sample. To detect the individual bands of molecules as they passed the detector, a laser with a suitable wavelength was focused to a single point in the microchannel, where it excited the fluorescent-labeled molecules. This type of setup (relatively simple channel layout, programmable power supplies to facilitate injection and electrophoretic separation, and laser-induced fluorescence) has been and still is the most commonly used microchip setup for analytical separation experiments. With this setup in place, other variants of electrophoresis can also be implemented on chips, such as gel electrophoresis and micellar electrokinetic chromatography (MEKC).

Let us briefly take a closer look at the most important modes of electrophoresis together with a few examples. General books and articles on capillary electrophoresis and related techniques are suggested for further reading [13–22], since in-depth treatment of these topics is far beyond the scope of this book. Covering just the many implementations of electrophoretic (and chromatographic) separation methods on microchips would require a book in itself. Some important techniques, such as isoelectric focusing (IEF) and isotachophoresis (ITP), are therefore

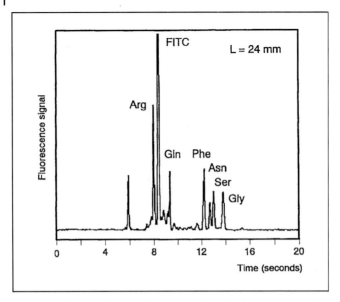

**Fig. 10.7** Early example of capillary electrophoretic separation on a chip. The analytes are fluorescent-labeled amino acids (reprinted in part with permission from [12]. Copyright (1993) American Chemical Society).

not described here, even though some microchip implementations have been described in journal articles [23–27] and reviews [28–33].

### 10.4.1
### Free-zone Electrophoresis

In free-zone electrophoresis the species to be separated are dissolved in buffer only (often called background electrolyte), i.e., the ions can freely move in the solution by diffusion and/or under the influence of an electric field. Every charged molecule can be characterized by its electrophoretic mobility, $\mu_{ep}$, which is largely determined by its size (radius, $r$) and charge, $q$:

$$\mu_{ep} = \frac{q}{6\pi\eta r} \tag{10.4}$$

where $\eta$ is the viscosity of the buffer system. When there is no electroosmotic flow the final velocity, $v$, of a charged molecule is derived as

$$v = v_{ep} = \mu_{ep} \cdot E \tag{10.5}$$

where $E$ is the electric field strength. However, most often, electroosmosis occurs as well, and then the final velocity is found as the vector sum of the electrophoretic velocity, $v_{ep}$, and the electroosmotic velocity, $v_{eo}$. Because the electroosmotic

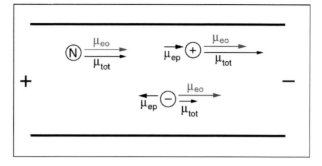

**Fig. 10.8** Vector addition of the electrophoretic mobilities of the individual ions and the electroosmotic mobility to yield total mobility.

flow, and hence the velocity associated with it, is a bulk property, it affects all molecules in the same way. Additionally, usually the absolute magnitude of the electroosmotic velocity is larger than any of the electrophoretic velocities, and therefore all species, positively charged, uncharged, or negatively charged, move in the same direction, namely the direction defined by the electroosmotic flow (Fig. 10.8).

Such an arrangement, with a sufficiently high electroosmotic flow, is one of the main reasons why capillary electrophoresis (CE) separations are so popular. A sample containing both cations and anions can be injected at one end of a capillary or a channel and then, after separation has occurred, the individual bands of ions can be detected at the other end. This makes the necessary experimental set-up much simpler. Separations in such systems are governed only by the charge-to-size ratio of the molecules. Therefore, small, highly charged cations are the first to migrate through the channel and arrive at the detector, followed by larger, less charged cations, followed by all the uncharged molecules, followed by larger, less charged anions. Finally the highly charged, small anions are the last to arrive at the detector. Because of the high separating power of CE, even small differences in charge-to-mass ratio can be sufficient to achieve separation. Good separations are generally characterized by sharp bands and short analysis times. In free zone CE, due to the absence of other interaction equilibria, and ignoring contributions from the injection plug width and the detector geometry, the only source of dispersion is diffusion, and the effects of diffusion can be minimized by decreasing the time from injection to detection. On microchips, extremely fast CE separations have been demonstrated, with analysis times below seconds and even below milliseconds [12, 34–36]. Fig. 10.9 shows the channel geometry around the injection cross and the main separation channel of a chip used for submillisecond separations. The channel widths were adjusted so that most of the voltage drop occurs in the short, narrow sections close to the cross, thus achieving very high field strengths, which according to Equation 10.5 lead to very high migration velocities.

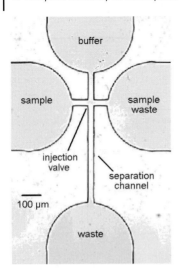

**Fig. 10.9** Channel geometry of a chip used for submillisecond CE separations. Only the region around the injection cross is shown (reprinted with permission from [36]. Copyright (1998) American Chemical Society).

A large number of free-zone electrophoretic separations have been performed on microchips, mainly due to the relative ease of implementation and the good separation efficiency, e.g., [30, 31].

## 10.4.2
## Gel Electrophoresis

For larger molecules, in particular DNA fragments, the charge-to-size ratio differences tend to become smaller and smaller, because many of these molecules consist of nearly identical units, and the addition or subtraction of units changes only the absolute charge and the absolute size, but not the ratio of those two parameters. To separate such molecules, the capillaries or channels are typically filled with a polymer gel that acts as a sieving matrix, constituting a molecular obstacle course that is negotiated faster by smaller molecules. The main transport mechanism is still based on migration in an electric field, however. A large variety of chemistries have been used for making suitable gels, but polyacrylamide-based gels are the most popular gels. When polymerizing the gels it is possible to adjust the pore size of the gels, thereby preparing gels for different separation tasks. Additionally, gels and coatings can help to suppress electroosmotic flow and to reduce adsorptive interactions with the channel walls. Unlike free-zone electrophoresis, electroosmotic flow is mostly undesired in gel electrophoresis, particularly for separation of negatively charged DNA fragments [37–39]. Fig. 10.10 shows separation of a number of DNA fragments (unknown peaks from a polymerase chain reaction (PCR) and marker molecules with a known number of base pairs) on a microchip, using a gel made of polymethylacrylamide. Although similar results can be obtained in conventional slab-gel electrophoresis, microchips have the potential of better separation efficiency, faster analysis times, easier handling, parallelization, and automation.

**Fig. 10.10** Gel electrophoretic separation of DNA fragments (from on-chip polymerase chain reactions (PCR) and marker fragments of a known number of base-pairs (reprinted with permission from [40]. Copyright (1998) American Chemical Society).

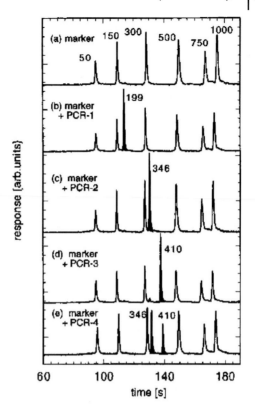

10.4.3
## Micellar Electrokinetic Chromatography (MEKC)

None of the techniques described above is suited to separating different neutral (uncharged) molecules, because (a) these molecules all have the same charge-to-size ratio in free zone CE (i.e., zero) and move with the velocity of the electroosmotic flow, and (b) they do not move at all in gel electrophoresis, where electroosmotic flow is suppressed. To separate neutrals it is necessary to introduce secondary equilibria, i.e., other interactions that can help distinguish molecules. These interactions can be hydrophobic interactions, guest–host interactions, or charge-transfer interactions. In MEKC, hydrophobic interactions are exploited. This is facilitated by adding a surfactant to the buffer solution. The surfactant, above a critical concentration, forms micelles, which are dynamic structures with a hydrophobic interior and a hydrophilic exterior. These micelles are typically highly charged anions displaying a high electrophoretic mobility opposed to the electroosmotic flow. In fact, these micelles often have a net velocity close to zero. Thus, the entirety of micelles within a buffer solution is referred to as a quasistationary phase. This phrase also implies that the separation mechanism is close to that of a chromatographic separation employing two phases.

MEKC is often regarded as a hybrid technique between pure CE and pure liquid chromatography. The separation mechanism is based on the different affinities of neutral molecules for the interior of the micelles. Molecules with higher hydrophobicity are more likely to be partitioned to a greater extent in the micellar phase and not so much in the buffer phase. However, we have to remember that these partitioning processes are dynamic. Whenever a neutral molecule is outside a micelle it is transported downstream by the electroosmotic flow. Whenever it is inside a micelle, it moves with the velocity of the micelle, i.e., hardly at all. Thus, highly hydrophobic molecules spend more time in the system before arriving at the detector than less-hydrophobic molecules.

A time window exists, defined by the fastest possible arrival time, i.e., the time corresponding to the velocity of the electroosmotic flow which is also the arrival time for a highly hydrophilic molecule never inside the micellar phase, and the slowest possible arrival time, i.e., the time it would take for a micelle to migrate from the injection point to the detector.

The principle ideas behind MEKC are summarized schematically in Fig. 10.11. One of the strengths of MEKC lies in the fact that many different surfactants can be used which vary in characteristics such as charge, hydrophobicity, and critical micellar concentration. The surfactant sodium dodecyl sulfate (SDS) is used most commonly. Other additives such as methanol, acetonitrile, or cyclodextrins can further modify the retention mechanism and allow a wide range of separation problems to be tackled. MEKC separations have been demonstrated on microchips with very good efficiencies (Fig. 10.12) and for different applications, including the analysis of explosives [41–43].

As mentioned, neutral compounds cannot be separated by electrophoresis. The need to separate neutral molecules triggered the invention of MEKC, which may,

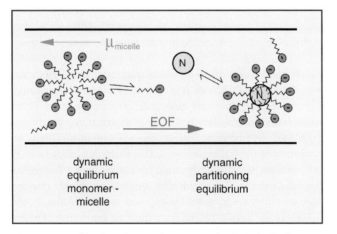

**Fig. 10.11** Micellar electrokinetic chromatography (MEKC). Surfactant molecules forming micelles and dynamic partitioning of neutral molecules between the micellar and buffer phases.

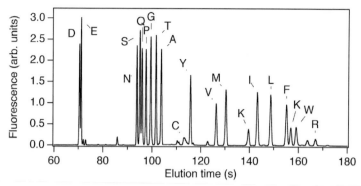

**Fig. 10.12** MEKC separation of 19 amino acids on a chip using a 11.87-cm-long separation channel (reprinted with permission from [43]. Copyright (2000) American Chemical Society).

however, be regarded as almost a chromatographic technique. Chromatographic separation itself is based on the presence of two phases, a stationary and a mobile phase, and the fact that different compounds display different affinities to those two phases. Many types of interactions can be exploited in chromatography, including hydrophobic, adsorption, charge-transfer, and size exclusion interactions. Chromatography works for charged and uncharged molecules, but is most often used for uncharged compounds. Again, within the scope of this book, we have to restrict ourselves discussing only a few of the many possible variants of chromatography.

One general distinction can be made according to whether the mobile phase is driven through the stationary phase hydrodynamically (e.g., by pumping) or by using the electroosmotic flow. Keeping in mind the discussion in Chapter 4 and what was mentioned in the beginning of this section, electroosmotic-based chromatographic implementations have predominated in the field of microchip analysis, mainly because of the better performance offered by the flat flow profile of EOF and also because of the non-availability of rugged, reliable, pulse-free miniature pumps.

Interestingly enough, one of the first analytical microchip devices was a gas chromatograph (where the mobile phase is a gas), which was designed and built in the mid-1970s [10]. However, the analytical chemistry community was not yet ready to embrace this new technology and further its development. Thus, not until the beginning of the 1990s did the first designs for a liquid chromatography (LC) chip (where the mobile phase is a liquid) appear [11]. In the remainder of this section we deal only with liquid chromatography chips, because LC is much more useable in the important field of separation of larger biomolecules and is also much easier to implement from a technological point of view than is gas chromatography on a chip. The biggest challenge to performing liquid chromatography on a chip is probably how to establish the stationary phases [29]. Four approaches are briefly explained and discussed in terms of their advantages and disadvantages.

10.4.4
**Open-channel Electrochromatography (OCEC)**

This is probably the method that is the easiest to implement. Using classical surface derivatization chemistry, octadecylsilane chains are immobilized on channel
walls, leaving most of the channel's lumen open. Because the species to be separated have to diffuse to the walls to interact with the stationary phase, the diffusion distances, i.e., the channel dimensions, have a crucial influence on the performance of these separation devices. Efficient separations have been demonstrated on microchips, with both isocratic and gradient elution. In isocratic elution, the buffer composition is not changed during the course of separation, and
in gradient elution the composition of the buffer (e.g., with respect to the amount
of organic modifier contents) is changed over time in a gradual or step-wise way.

Gradient elution (also referred to as solvent programming) allows more fine-
tuned optimization of complex separation problems. The implementation of solvent programming on chips is facilitated by the excellent fluidic control offered by

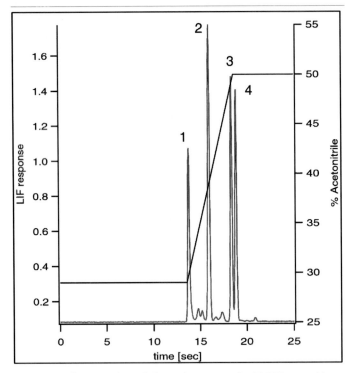

**Fig. 10.13** A fast open-channel electrochromatography (OCEC) on a chip
utilizing a gradient elution technique, by which the concentration of acetonitrile in the buffer is increased linearly from 29% to 50% within 5 s
(reprinted with permission from [44]. Copyright (1998) American Chemical
Society).

electroosmotic flow. An example of a fast OCEC separation of four neutral dyes using a linear solvent gradient is shown in Fig. 10.13 [44]. The biggest disadvantage of OCEC is the limited loadability, i.e., the small amount of absolute mass that can be injected. Because the stationary phase consists of only a thin layer on the channel walls, it has a very limited number of interaction sites. These are easily saturated by too large an amount of injected sample, which in turn drastically degrades the separation efficiency. On the other hand, if only small amounts of sample can be injected, this poses a bigger challenge for the detectability of the separated bands of molecules.

## 10.4.5
### Packed-bed Chromatography

Classically, liquid chromatography has been performed in columns packed with micrometer-sized porous particles, which can be chemically derivatized and which constituted the stationary phase through which the mobile phase percolated. Efficiency in such systems depends strongly on the quality of the packed bed, i.e., how uniform the particles are and how densely and regularly they are packed. Therefore, particles were often packed under high pressure. Additionally, the particles had to be kept inside the column or capillary by some means, usually by porous sintered glass frits at the inlet and outlet. Planar chips with bonded lids are not as mechanically stable as cylindrical capillaries, and therefore neither high-pressure packing methods nor high-pressure operating conditions are typically useable for microchips. However, packing microchannels with 1.5–4-μm particles with the help of electric fields yields nicely packed beds [45], which are held in place by micromachined weir structures, which leave only a small gap between the raised channel floor and the cover plate. This gap is large enough for liquid to pass through but not large enough to allow particles to escape from the bed (Fig. 10.14).

Another way to get around the problem of having to have frits inside a channel is to use tapered channels. As particles pass through a tapered section, they have less and less room, running into each other, and eventually creating a blockage. Behind this blockage the chromatographic bed itself is then built up. Even though the taper blocks the passage of particles, it is still permeable to the liquid phase, thus enabling chromatographic separation. Creating a mechanically stable blockage structure in this way resembles the construction of a stable arch with a keystone – this principle is schematically depicted in Fig. 10.15 [46].

## 10.4.6
### Microfabricated Stationary-phase Support Structures

Above, we mentioned that the quality of separation with a packed bed depends on the size uniformity of the particles making up the bed as well as on the geometric uniformity of packing. Microfabrication now offers the possibility to micromachine a 'packed' chromatographic bed by using a photolithographic process that leaves

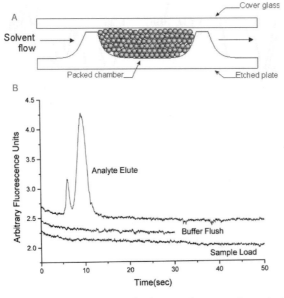

**Fig. 10.14** A) Cross section of a device with weirs and a packed chromatographic bed. B) A simple electrochromatographic separation after enrichment of sample on the same phase material (see also section 10.5.1) (reprinted with permission from [45]. Copyright (2000) American Chemical Society).

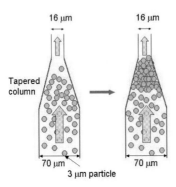

**Fig. 10.15** Schematic of the keystone effect, which prevents particles from entering the narrow channel section, thus forming a barrier for packing the stationary phase (reprinted with permission from [46]. Copyright (2002) American Chemical Society).

well-defined, equally spaced, and regularly arranged posts in the flow path [47–49]. These posts are the particles of a geometrically very-well-defined chromatographic stationary phase support. The stationary phase itself is again, as with real silica particles, chemically bound to the surfaces of these posts. By using photolithography and etching techniques, the characteristics of the stationary phase support, and consequently the chromatographic performance of the final microchip system, can be easily adjusted. Parameters such as 'particle' size, 'particle' shape, spacing, and spa-

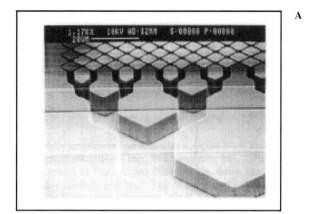

A

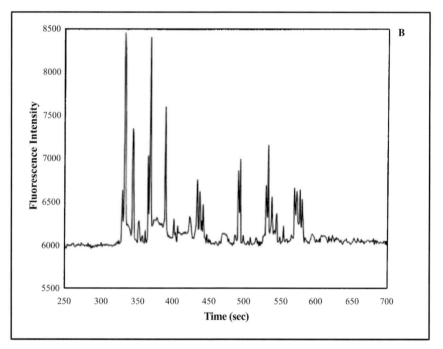

B

**Fig. 10.16** (A) Microphotograph of the entrance to a micromachined regular bed for chromatography; (B) a separation of peptides from an ovalbumin digest (reprinted in part with permission from [47, 49], respectively. Copyright (1998) American Chemical Society and (1999) Elsevier).

tial orientation can be changed by using different etch masks, and the effect on the separation performance can be investigated. Fig. 10.16a shows a layout with fluidic splitters and the beginning of a chromatographic support bed with diamond-shaped posts. After surface modification with octadecylsilane, excellent separations were demonstrated on such a microchip system (Fig. 10.16b).

10.4.7
**In-situ-polymerized Stationary Phases**

Probably the most versatile approach to creating stationary phases in microchips designed for chromatography involves the use of monomer solutions that can be filled into the channels and subsequently polymerized in situ. Ideally, a more-or-less porous solid rod (or monolith) is formed, which either already contains the interaction sites necessary for chromatographic separations or is easily modified in a following step. The great advantages of this approach are that (a) the monomer solution usually has low viscosity and is therefore easily filled into the channels without the need for high pressure, and (b) the composition of the monomer solution can be tailored to match a certain set of requirements. The monomer cocktail can, for example, contain functional entities to assure that the desired chromatographic interaction will occur. Other functional entities (e.g., sulfonate groups) can help ensure a strong enough electroosmotic flow, and a third group of additives, the so-called porogens, assure that the final porous polymer has pores of a certain average size. This is important for tailoring the fluidic resistance of such a chromatographic column.

An additional benefit of in-situ polymerization is the fact that no end frits or other means to keep the bed inside the channel are necessary, because the monolithic rod is usually anchored to the channel walls during polymerization. Several different chemistries can be used to create porous polymers suitable for use as chromatographic supports or phases inside microchannels. The three most often used chemistries are (a) polyacrylamides or methacrylates [50–52], (b) styrene–divinylbenzene copolymers [53, 54], and (c) silicates (sol-gels) [55, 56].

Polymerization can be induced thermally by heating the monomer mixture or photochemically by irradiating the mixture with light of an appropriate wavelength. Irradiation offers the great advantage of selective polymerization, i.e., polymerization can be prevented in places that are shielded from the light. Thus, the use of masks enables defining sections of the channel network where no polymerization occurs. This is often advantageous in the injection and detection areas, where a stationary phase might interfere with proper functioning. On the other hand, we have to be aware of the fact that, at interfaces between a polymerized phase and the free channel, parameters such as the electroosmotic flow change abruptly. Such changes can induce (mostly undesired) effects, which can contribute significantly to dispersive band broadening. Fig. 10.17a shows several parts of a microfluidic chip, in which a photodefined polymeric monolith was used for separating peptides (Fig. 10.17b). The interface between the monolith and the free channel can be seen, as well as the injection and detection areas, which are free of polymer [57].

So far in this section we have mainly looked at separation methods that are already established in more traditional formats (columns, capillaries, slab gels, etc.) and have been implemented on microchips. Most of these techniques have now been optimized to the point where they perform as well and often better on microchips. There are, however, also variants of these fundamental techniques that

**Fig. 10.17** **A** Injection region (a), detection region (b), and interface (c) between a polymerized part of the channel and an open part; **B** separation of six peptides on a photopolymerized stationary phase (reprinted with permission from [57]. Copyright (2002) American Chemical Society).

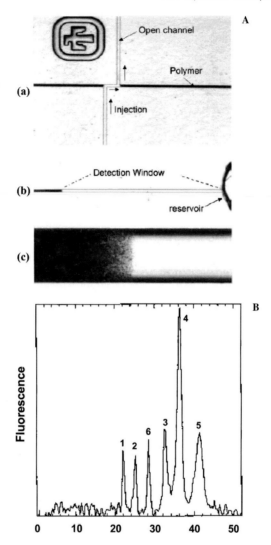

can be implemented only in a miniaturized format. In the remainder of this section we briefly look at four interesting examples.

10.4.8
**Synchronous Cyclic Capillary Electrophoresis (SCCE)**

Fig. 10.18 shows the layout of a chip featuring a continuous separation 'loop' and several side channels allowing fluidic and electrical contact to the central loop [58]. After injection of a sample plug at the cross intersection, voltage is applied so that electrokinetic movement occurs through one half of the loop (from elec-

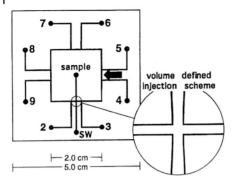

**Fig. 10.18** Channel layout of a chip used for synchronous cyclic capillary electrophoresis (SCCE) (reprinted with permission from [58]. Copyright (1996) American Chemical Society).

trode 2 to electrode 6 in Fig. 10.18). After the bands have traveled a certain distance, the electric field is applied between two other electrodes (electrode 4 and electrode 8) so that the field switching is in synchrony with the movement of the bands to ensure continuous cyclic movement. In this manner, a theoretically infinite separation length is available. However, there are of course some restrictions.

As the separation of bands moving at different velocities progresses, bands that are too fast or too slow lose synchrony with the field switching. Consequently these bands are either pushed out of the loop toward a waste reservoir or are left behind when a switch occurs. In other words, synchronization only works for a specific band of molecules with a specific electrokinetic velocity [59]. Thus, this method is useful for separating species having very similar mobilities, which require a long separation length before they can be separated.

The other restriction is given by dispersion of the bands. Even if diffusion is assumed to be the only source for dispersion, bands moving through the loop eventually become so dispersed that they fill the entire volume of the loop, thereby making further separation impossible. Implementation of SCCE in a more traditional format is not impossible but is very cumbersome. In particular, the required capillary coupling is difficult to achieve [60, 61]. In a planar chip format this is comparatively easy.

10.4.9
**Two-dimensional Separations**

Separations involving one particular separation mechanism are also denoted one-dimensional separations and can separate only a limited number of components. Increasing the available separation length does not significantly increase the number of peaks that can be resolved, but does increase the analysis time. The number of peaks that can be resolved by a separation system is the peak capacity. To truly increase peak capacity, more than one separation method needs to be employed. These methods are then coupled to one another, arriving at two- or more-dimensional separations.

**Fig. 10.19** Chip layout for 2D separation (*top*). Separation in the first dimension (OCEC) was performed in a 25-cm-long spiral channel, and in the second dimension (CE) in a 1.2-cm-long channel. Results from the separation of a complex peptide mixture are shown *on the bottom* (reprinted with permission from [63]. Copyright (2001) American Chemical Society).

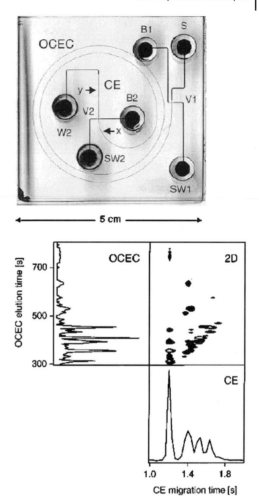

Separation in two dimensions can be coupled in a *parallel fashion*, meaning that the separation in the first dimension is run until the fastest components arrive at a detector. Subsequently the entire contents of the channel are injected into an array of parallel channels turned 90° with respect to the original channel.

Alternatively, two separation systems can be coupled in a *serial manner*, by which 'chunks' coming off the first dimension are continuously injected into a (rather fast) second dimension. For 2D separations to work optimally, i.e., to achieve maximum peak capacity, the two separation mechanisms should be as different as possible from each other: they should be orthogonal to each other with respect to the principle by which they separate components [62]. For example, a chromatographic technique can be used in the first dimension and an electrophoretic technique in the second.

A chip for serial coupling of OCEC and CE is shown in Fig. 10.19, together with a 2D plot of the separation of a complex sample (a peptide mixture) [63]. Isoelectric focusing can also be coupled with free-zone electrophoresis [24, 64]. Again, planar microchips have several advantages in creating 2D separation systems: on the one hand, coupling of different separation systems does not require carefully designed connectors, but is accomplished by proper design of the channel layout; on the other hand, for efficient 2D separations a very fast second dimension is essential.

As we have seen, extremely fast CE separations have been demonstrated on microchips. The chip shown in Fig. 10.19 uses a channel only 1.2 cm long for the CE dimension, thereby achieving fast, but still very efficient separation.

## 10.4.10
### Hydrodynamic Chromatography (HDC)

This is a separation method for particles or larger molecules (DNA molecules or polymer molecules) that differ in size. Despite its name, it is not truly a chromatographic method, since it does not require a secondary interaction for the separation. It does, however, require very shallow channels (a few micrometers or even less), because it uses the velocity differences in different parts of a microchannel as given by a hydrodynamic flow profile.

Smaller molecules can sample a large variety of velocities, because they can get very close to the channel walls. Larger molecules, on the other hand, cannot come closer to the channel walls than the distance determined by their radius, and they consequently experience mostly the faster parts of the hydrodynamic flow profile (Fig. 10.20). As a result, larger molecules have, on average, a higher velocity than smaller molecules, resulting in separation based on size.

Obviously, in too-large channels this effect is not very pronounced, which is why rather shallow channels have to be used. HDC on microchips having channels 1 μm deep and 1 mm wide was demonstrated recently [65, 66]. An aspect ratio of 1/1000 was chosen to allow optical detection along the long axis. Fabricating channels with such aspect ratios is a challenge and really only feasible by micro-

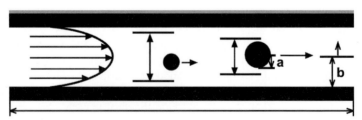

**Fig. 10.20** Principle of hydrodynamic chromatography (see text for details) (reprinted with permission from [66]. Copyright (2002) American Chemical Society).

technology. Larger structures with such aspect ratios are not stable, i.e., the roof would need to be supported to prevent it from collapsing.

10.4.11
## Shear-driven Chromatography

Instead of relying on a difference in pressure or voltage to support flow, in this technique the bottom part of a microfluidic channel is held stationary, while the top part is moved at a constant velocity (Fig. 10.21). This leads to the development of a flow profile in which the highest velocity is at the top and the lowest velocity at the bottom. Since the bottom part is held stationary, the lowest velocity is zero, and the average velocity in the center equals half the maximum velocity. Such a flow is called shear-driven flow or Couette flow.

Although technically more challenging to implement, shear-driven flows have been used for chromatographic separation in microsystems in which the stationary wall is chemically modified with a chromatographic phase material. In hydrodynamically-driven separation systems it is necessary to reduce the channel dimensions or the particle dimensions to increase the overall separation performance. However, this comes at the price of increased counterpressure, which at some point becomes prohibitively high and limits further optimization. Shear-driven chromatography is not limited by counterpressure, and theoretically the performance can be increased indefinitely by using smaller and smaller gaps between the stationary and moving walls. Still, going to smaller dimensions means much smaller volumes, and at some point the injection and detection steps become limiting factors again.

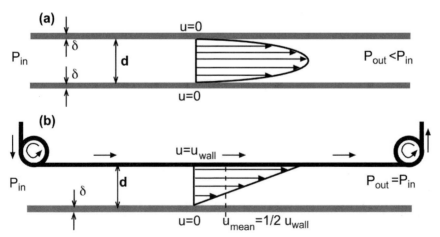

**Fig. 10.21** Hydrodynamic (pressure-driven) flow (a) and shear-driven flow (b). Note that no pressure drop occurs in shear-driven flow (reprinted with permission from [67]. Copyright (2000) American Chemical Society).

**10.5**
**Other Analytical Techniques**

Having now delved into a discussion of separation systems in more detail, we continue this section with a discussion of additional system parts. Separation systems, although often constituting the heart of a μTAS system, rarely stand all alone. Around them, on the front or on the back, other units are coupled to the separation units. The tasks of these other units are often summarized under the term *sample preparation* or *sample treatment* and can involve many different procedures. The following list, though incomplete, gives an idea of the many possibilities to treat samples before they go through a separation or detection step:

– concentration and enrichment
– filtration and washing
– extraction, transition from one phase to another
– mixing and dilution
– staining and labeling (e.g., for detection purposes)
– cell handling (counting/sorting/isolation/lysing)
– DNA purification, amplification, PCR, etc.
– enzyme digestion
– flow sheathing for nebulization/electrospray ionization

Again, many examples are available in the literature of how these functionalities have been implemented on microchips, and we have to restrict ourselves to a few examples. For a very comprehensive recent review on sample preparation techniques on microfabricated devices see [68]. Integrating several of these functionalities onto one microchip is still one of the main challenges on the road to true micrototal analysis systems.

10.5.1
**Solid-phase Extraction (SPE)**

Stationary phases for chromatography are not limited to separation purposes, they can also help to enrich or concentrate interesting species from a diluted sample solution. To achieve this, a mostly aqueous buffer solution containing the analytes of interest is continuously percolated through the stationary phase (remember, in separations we only inject a small plug of sample). Under the appropriate conditions, the analytes interact strongly with the stationary phase and are held back. The more solution we pump through the stationary phase, the more analyte molecules are held back, until a point at which the capacity of the stationary phase is reached. At this point, we can switch to a new buffer solution containing a much higher concentration of an organic modifier (e.g., methanol or acetonitrile). Once we pump this solution through the stationary phase, all analyte molecules are washed off ('eluted') and collected, in a much smaller volume than before, thus increasing their concentration. Before elution, washing steps can be included,

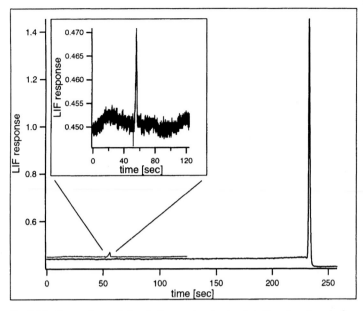

**Fig. 10.22** Comparison of signals from a simple plug injection (inset) and after a solid-phase extraction step (reprinted with permission from [69]. Copyright (2000) John Wiley & Sons, Inc.).

which can help get rid of other components of the original sample solution, components that might interfere with the final detection step.

SPE has been implemented on chips using the same devices as for chromatographic separations. One example was already mentioned in connection with packed-bed chromatography (Fig. 10.14b) [45]. This figure shows three traces, the lowest of which is the signal obtained during the enrichment step. Because only a baseline signal is evident, this means that all analyte molecules were retained on the stationary phase. The second trace, pertaining to the washing step, is also only a baseline signal. The third trace shows the elution step, in which the analytes are washed off and (in this case) also separated, yielding two strong signals.

Results from a similar experiment with an OCEC chip are depicted in Fig. 10.22 [69]. Here, the inset shows the minute signal available when injecting a small plug of analytes at the given concentration. After enrichment and elution, however, the signal is much larger. Here, an 8.7 nM solution of a neutral dye was enriched for 160 s, corresponding to an extracted volume of 30.4 nL and an absolute amount of 0.27 fmol. Elution was possible in only 0.57 nL, yielding a final concentration of 750 nM, assuming everything retained was washed off the stationary phase. This in turn corresponds to an enrichment factor of about 86. With the advent of tailor-made in-situ polymerized stationary phases, much higher enrichment factors (up to 1000) have been reported more recently [70].

10.5.2
## Electrokinetic Enrichment of DNA

An interesting device design has been published, by which DNA molecules were enriched at a porous junction between two channels [71] (Fig. 10.23, top). Two adjacent channels are separated by a small gap of 3–12 μm. The width of the gap isolating the two channel segments is controlled by the mask design and the etching conditions. When bonding a cover onto this structure a technique was used that involved a thin layer of silicate spun onto the top wafer. After bonding, this resulted in a thin porous layer between the bottom and top layers. Due to the close proximity of the two channel segments, the porous layer connecting them can act like a salt bridge and allow electrical contact through the porous layer from one channel segment to the other.

Clearly, though, the porous layer also constitutes a physical barrier allowing only small charge carriers to pass through. If an electric field is applied to electrophoretically transport DNA molecules from the left part of the channel towards the junction (Fig. 10.23, bottom), these molecules cannot penetrate the porous layer and are hence concentrated in front of it, while the electrical connection is maintained through the junction. After a certain time, when the desired amount of DNA has been enriched, or when the enriched plug of DNA molecules becomes too large, the voltages are switched to inject the enriched plug into the main channel for, e.g., separation.

Besides using porous layers, similar junctions have been made by having two channels close together and applying high field strengths until electrical breakthrough occurs, thus creating a structural defect in the material separating the two channels and defining a preferred pathway for the current.

10.5.3
## Electrostacking

As a final example, electrostacking methods can be used to concentrate dilute samples. Electrostacking includes several techniques that use differences in migration behavior in spatially defined regions of different field strengths. A good overview of such techniques on chips can be found in [68].

Regions with different field strengths can be created by using zones of solutions having different conductivities. For instance, if the sample is dissolved in a buffer with a lower conductivity than the background buffer, then the electric field strength is higher in this sample plug. Charged ions then have a higher migration velocity inside the plug while on their way toward the respective electrode. However, as they arrive at the border between the low-conductivity sample plug and the high-conductivity background buffer, they experience a sharp drop in the electric field and an accompanying sharp drop in their migration velocities. This is not unlike 'running into a brick wall' and results in concentration of the charged sample molecules at the borders of the original sample plug.

**Fig. 10.23** Schematic view of a porous junction structure (*top*) and a sequence showing the enrichment of fluorescently labeled DNA and its subsequent injection (*bottom*) (reprinted with permission from [71]. Copyright (1999) American Chemical Society).

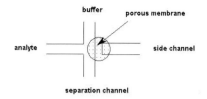

**Top view**

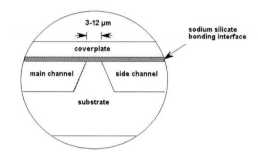

**Cross section view**

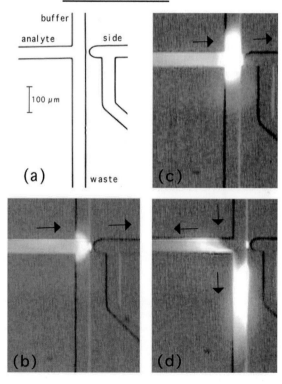

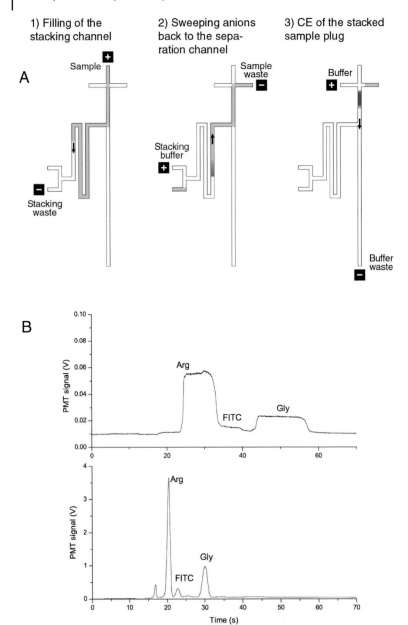

**Fig. 10.24** (a) Principle of large-volume electrostacking with a dedicated side channel; (b) comparison of large-volume injection, unstacked (upper trace) and stacked (lower trace) (reprinted with permission from [72]. Copyright (2001) Wiley-VCH).

A chip design using a side channel to enrich ions from the entire volume of the side channel is shown in Fig. 10.24a [72]. First, the side channel is filled with the dilute sample solution. Then, by a proper combination of electrophoretic and electroosmotic transport, the sample solution is slowly moved back towards the main channel intersection, followed by a plug of high-conductance buffer. This leads to stacking and enrichment of the negatively charged sample compounds at the back end of the original sample plug. When everything in the volume defined by the side channel has been stacked, this concentrated plug can be injected into the main channel for electrophoretic separation of its components.

Fig. 10.24b shows the results obtained for separation after injection of a stacked sample, as compared to separation when the same original volume was injected directly, i.e., without stacking. Direct injection of large volumes is clearly not an option because it leads to volume overloading and consequently bad peak shapes. On the other hand, the injection of a large volume that has been stacked leads to good peak shapes and improved signal-to-noise ratios.

## 10.6
## References

1 KELLNER, R.; MERMET, J.-M.; OTTO, M.; WIDMER, H.M. *Analytical Chemistry*, Wiley-VCH: Weinheim, **1998**.

2 MANZ, A.; GRABER, N.; WIDMER, H.M. *Sensors and Actuators B* **1990**, 244–248.

3 REYES, D. R.; IOSSIFIDIS, D.; AUROUX, P. A.; MANZ, A. *Analytical Chemistry* **2002**, 74, 2623–2636.

4 AUROUX, P.-A.; IOSSIFIDIS, D.; REYES, D. R.; MANZ, A. *Analytical Chemistry* **2002**, 74, 2637–2652.

5 RAMSEY, J. M. *Nature Biotechnology* **1999**, 17, 1061–1062.

6 RUZICKA, J.; HANSEN, E. H. *Anal. Chim. Acta* **1975**, 78, 145.

7 RUZICKA, J.; HANSEN, E. H. *Flow Injection Analysis*, 2nd ed., **1988**.

8 KROG, J. P.; DIRAC, H.; FABIUS, B.; GRAVESEN, P.; DARIDON, A.; LICHTENBERG, J.; VERPOORTE, E.M.J.; DE ROOIJ, N. F.; PENNARUN-THOMAS, G.; SEQUEIRA, M.; DIAMOND, D.; DENNINGER, M.; GESCHKE, O.; KUTTER, J. P.; HOWITZ, S.; STREC, C.; CHARLES, P.; COGNET, L., Enschede, Netherlands, May 14–18, **2000**; Kluwer: Amsterdam, 419–422.

9 DARIDON, A.; SEQUEIRA, M.; PENNARUN-THOMAS, G.; DIRAC, H.; KROG, J. P.; GRAVESEN, P.; LICHTENBERG, J.; DIA-

MOND, D.; VERPOORTE, E.; DE ROOIJ, N. F. *Sensors and Actuators B* **2001**, 76, 235–243.

10 TERRY, S. C.; JERMAN, J. H.; ANGEL, J. B. *IEEE Trans. Electron. Devices* **1979**, 26, 1880–1887.

11 MANZ, A.; MIYAHARA, Y.; MIURA, J.; WATANABE, Y.; MIYAGI, H.; SATO, K. *Sensors and Actuators B* **1990**, 1, 249–255.

12 EFFENHAUSER, C. S.; MANZ, A.; WIDMER, H.M. *Analytical Chemistry* **1993**, 65, 2637–2642.

13 JORGENSON, J.W.; LUKACS, K.D. *Analytical Chemistry* **1981**, 53, 1298–1302.

14 GIDDINGS, J.C. *Unified Separation Science*; Wiley: New York, **1991**.

15 FOLEY, J.P. *Analytical Chemistry* **1990**, 62, 1302–1308.

16 SEPANIAK, M.J.; POWELL, A.C.; SWAILE, D.F.; COLE, R.O. In *Capillary Electrophoresis: Theory and Practice*; GROSSMAN, P. D., COLBURN, J.C., Eds.; Academic Press.: San Diego, 1992, pp 159–189.

17 RIGHETTI, P.G. In *Isoelectric Focusing: Theory, Methodology and Applications*; Elsevier:Amsterdam, **1983**.

18 CREGO, A.L.; GONZALEZ, A.; MARINA, M.L. *Critical Reviews in Analytical Chemistry* **1996**, 26, 261–304.

19  GROSSMAN, P.D.; COLBURN, J.C. *Capillary Electrophoresis: Theory and Practice*; Academic Press: San Diego, **1992**.

20  MANZ, A.; HARRISON, D.J.; VERPOORTE, E.M.J.; FETTINGER, J.C.; PAULUS, A.; LUDI, H.; WIDMER, H.M. *Journal of Chromatography* **1992**, *593*, 253–258.

21  POPPE, H. In *Advances in Chromatography*, **1998**; *38*, 233–300.

22  LI, S.F.Y. *Capillary Electrophoresis*; Elsevier: Amsterdam, **1993**.

23  HOFMANN, O.; CHE, D.; CRUICKSHANK, K.A.; MÜLLER, U.R. *Analytical Chemistry* **1999**, *71*, 678–686.

24  HERR, A.E.; MOLHO, J.I.; BHARADWAJ, R.; MIKKELSEN, J.C.; SANTIAGO, J.G.; KENNY, T.W.; BORKHOLDER, D.A.; NORTHRUP, M.A., Monterey, California, USA **2001**; Kluwer: Amsterdam, 51–53.

25  RAISI, F.; BELGRADER, P.; BORKHOLDER, D.A.; HERR, A.E.; KINTZ, G.J.; POURHAMADI, F.; TAYLOR, M.T.; NORTHRUP, M.A. *Electrophoresis* **2001**, *22*, 2291–2295.

26  CASLAVSKA, J.; THORMANN, W. *Journal of Microcolumn Separations* **2001**, *13*, 69–83.

27  MASAR, M.; ZUBOROVA, M.; BIELCIKOVA, J.; KANIANSKY, D.; JOHNCK, M.; STANISLAWSKI, B. *Journal of Chromatography A* **2001**, *916*, 101–111.

28  EFFENHAUSER, C.S. In *Microsystem Technology in Chemistry and Life Science*; BECKER, H., MANZ, A., Eds., **1998**; Vol. 194, pp 51–82.

29  KUTTER, J. P. *TrAC – Trends in Analytical Chemistry* **2000**, *19*, 352–363.

30  BRUIN, G.J.M. *Electrophoresis* **2000**, *21*, 3931–3951.

31  DOLNIK, V.; LIU, S.; JOVANOVICH, S. *Electrophoresis* **2000**, *21*, 41–54.

32  GEBAUER, P.; BOCEK, P. *Electrophoresis* **2002**, *23*, 3858–3864.

33  SHIMURA, K. *Electrophoresis* **2002**, *23*, 3847–3857.

34  HARRISON, D.J.; FAN, Z.H.; SEILER, K.; MANZ, A.; WIDMER, H.M. *Analytica Chimica Acta* **1993**, *283*, 361–366.

35  JACOBSON, S.C.; HERGENRÖDER, R.; KOUTNY, L.B.; RAMSEY, J.M. *Analytical Chemistry* **1994**, *66*, 1114–1118.

36  JACOBSON, S.C.; CULBERTSON, C.T.; DALER, J.E.; RAMSEY, J.M. *Analytical Chemistry* **1998**, *70*, 3476–3480.

37  ULFELDER, K.J.; MCCORD, B., R. In *Handbook of Capillary Electrophoresis*; LANDERS, J.P., Ed.; CRC Press: Boca Raton, FL, **1997**.

38  VERPOORTE, E.M.J. *Electrophoresis* **2002**, *23*, 677–712.

39  EHRLICH, D.J.; MATSUDAIRA, P. *Trends in Biotechnology* **1999**, *17*, 315–319.

40  WATERS, L.C.; JACOBSON, S.C.; KROUTCHININA, N.; KHANDURINA, J.; FOOTE, R.S.; RAMSEY, J.M. *Analytical Chemistry* **1998**, *70*, 5172–5176.

41  KUTTER, J.P.; JACOBSON, S.C.; RAMSEY, J.M. *Analytical Chemistry* **1997**, *69*, 5165–5171.

42  WALLENBORG, S.R.; BAILEY, C.G. *Analytical Chemistry* **2000**, *72*, 1872–1878.

43  CULBERTSON, C.T.; JACOBSON, S.C.; RAMSEY, J.M. *Analytical Chemistry* **2000**, *72*, 5814–5819.

44  KUTTER, J.P.; JACOBSON, S.C.; MATSUBARA, N.; RAMSEY, J.M. *Analytical Chemistry* **1998**, *70*, 3291–3297.

45  OLESCHUK, R.D.; SHULTZ-LOCKYEAR, L.L.; NING, Y.; HARRISON, D.J. *Analytical Chemistry* **2000**, *72*, 585–590.

46  CERIOTTI, L.; DE ROOIJ, N.F.; VERPOORTE, E. *Analytical Chemistry* **2002**, *74*, 639–647.

47  HE, B.; TAIT, N.; REGNIER, F.E. *Analytical Chemistry* **1998**, *70*, 3790–3797.

48  HE, B.; REGNIER, F.E. *Journal of Pharmaceutical and Biomedical Analysis* **1998**, *17*, 925–932.

49  HE, B.; JI, J.; REGNIER, F.E. *Journal of Chromatography A* **1999**, *853*, 257–262.

50  LIAO, J.L.; CHEN, N.; ERICSON, C.; HJERTEN, S. *Analytical Chemistry* **1996**, *68*, 3468–3472.

51  PETERS, E.C.; PETRO, M.; SVEC, F.; FRECHET, J.M.J. *Analytical Chemistry* **1998**, *70*, 2288–2295.

52  PETERS, E.C.; PETRO, M.; SVEC, F.; FRECHET, J.M.J. *Analytical Chemistry* **1998**, *70*, 2296–2302.

53  WANG, Q.C.; SVEC, F.; FRECHET, J.M.J. *Analytical Chemistry* **1993**, *65*, 2243–2248.

54  GUSEV, I.; HUANG, X.; HORVATH, C. *Journal of Chromatography A* **1999**, *855*, 273–290.

55  DULAY, M.T.; KULKARNI, R.P.; ZARE, R.N. *Analytical Chemistry* **1998**, *70*, 5103–5107.

**56** TANG, Q. L.; XIN, B. M.; LEE, M. L. *Journal of Chromatography A* **1999**, *837*, 35–50.

**57** THROCKMORTON, D. J.; SHEPODD, T. J.; SINGH, A. K. *Analytical Chemistry* **2002**, *74*, 784–789.

**58** VON HEEREN, F.; VERPOORTE, E.; MANZ, A.; THORMANN, W. *Analytical Chemistry* **1996**, *68*, 2044–2053.

**59** MANZ, A.; BOUSSE, L.; CHOW, A.; METHA, T. B.; KOPF-SILL, A.; PARCE, J. W. *Fresenius Journal of Analytical Chemistry* **2001**, *371*, 195–201.

**60** ZHAO, J. G.; HOOKER, T.; JORGENSON, J. W. *Journal of Microcolumn Separations* **1999**, *11*, 431–437.

**61** ZHAO, J. G.; JORGENSON, J. W. *Journal of Microcolumn Separations* **1999**, *11*, 439–449.

**62** ROCKLIN, R. D.; RAMSEY, R. S.; RAMSEY, J. M. *Analytical Chemistry* **2000**, *72*, 5244-5249.

**63** GOTTSCHLICH, N.; JACOBSON, S. C.; CULBERTSON, C. T.; RAMSEY, J. M. *Analytical Chemistry* **2001**, *73*, 2669–2674.

**64** HERR, A. E.; MOLHO, J. I.; DROUVALAKIS, K. A.; MIKKELSEN, J. C.; UTZ, P. J.; SANTIAGO, J. G.; KENNY, T. W. *Analytical Chemistry* **2003**, *75*, 1180–1187.

**65** BLOM, M. T.; CHMELA, E.; GARDENIERS, J. G. E.; TIJSSEN, R.; ELWENSPOEK, M.; VAN DEN BERG, A. *Sensors and Actuators B-Chemical* **2002**, *82*, 111–116.

**66** CHMELA, E.; TIJSSEN, R.; BLOM, M. T.; GARDENIERS, H. J. G. E.; VAN DEN BERG, A. *Analytical Chemistry* **2002**, *74*, 3470–3475.

**67** DESMET, G.; BARON, G. V. *Analytical Chemistry* **2000**, *72*, 2160–2165.

**68** LICHTENBERG, J.; DE ROOIJ, N. F.; VERPOORTE, E. *Talanta* **2002**, *56*, 233–266.

**69** KUTTER, J. P.; JACOBSON, S. C.; RAMSEY, J. M. *Journal of Microcolumn Separations* **2000**, *12*, 93–97.

**70** YU, C.; DAVEY, M. H.; SVEC, F.; FRECHET, J. M. J. *Analytical Chemistry* **2001**, *73*, 5088–5096.

**71** KHANDURINA, J.; JACOBSON, S. C.; WATERS, L. C.; FOOTE, R. S.; RAMSEY, J. M. *Analytical Chemistry* **1999**, *71*, 1815–1819.

**72** LICHTENBERG, J.; VERPOORTE, E.; DE ROOIJ, N. F. *Electrophoresis* **2001**, *22*, 258–271.

**73** CLARK, L. C.; *Trans. Am. Soc. Art. Intern. Organs.* **1956**, *2*, 41–48.

**74** GESCHKE, O., Ph.D. Dissertation, University of Muenster, Germany, **1998**.

**75** CAMMANN, K., *Instrumentelle Analytische Chemie*, **2000**, Spektrum Akademischer Verlag, Heidelberg.

**76** KOOIJMAN, S. A. L., *Water Research* **1996**, *30*, 1711–1723.

**77** MEYER, H. et al, *Sensors and Actuators B* **1994**, *18*, 229–234.

# Subject Index

*Microsystem Engineering of Lab-on-a-chip Devices*
O. Geschke, H. Klank, P. Telleman
Copyright © 2004 Wiley-VCH Verlag GmbH & Co. KGaA, Weinheim
ISBN: 3-527-30733-8

– additives 161
– biocompatibility 161
– borofloat 162
– dry chemical 163
– laser micromachining 163
– light sensitive glass 161
– mechanical stability 161
– micromachining 162
– wet etching 163
glass transition temperature 161, 165, 170, 171
gloves 11, 163
glucose 224
– level 220
– oxidase 219
gluing 176
gold 33, 74, 139, 162, 176
Gouy-Chapman layer 46
grid 84, 86, 103, 108
– dimensions 97
– generation 58, 84, 86, 90
– hybrid 86
– orthogonality 99
– refinement 88
– resolution 97
– skewed 95
– solution independency 88
– structured 84, 86, 89, 92, 169
– unstructured 84, 86, 89, 92

**h**

Hagen-Poiseuille 18, 96
heat convection 21, 25
heater 60, 59, 120
HEPA-filter 9
HF *see* hydrofluoric acid
H-filter 28
High Efficiency Particulate Air filtering *see* HEPA-filter
high throughput screening 3
HNA 143
hot embossing 172, 173
HTS *see* high throughput screening
hydrochloric acid 33
hydrofluoric acid 33, 133, 141 ff, 156, 163
hydrogels 41
– pH sensitive 41
hydrogen peroxide 160, 220, 222
hydrophobic patch 45, 54

**i**

identification of components 213
immune system 219
ingot 119

injecting 54
injection molding 134, 171, 173
integrated circuit 1, 2, 64
interconnections 185, 189, 194, 205, 249
– electrical interconnections 183, 184, 190, 195, 206
– fluidic interconnections 191, 195, 201 ff
– optical interconnections 195, 201, 206
Ion Selective Field Effect Transistor *see* ISFET
ionophores 218
– valinomycine 218
ISFET 74, 75, 218
– pH-ISFET 75
– ReFET 76
isoelectric focusing 240
isotropic etching 127, 141, 163

**j**

Joule heating 83

**k**

kalomel 74
kinematic viscosity 17
Kirchhoff rules 54

**l**

lag 149
Lambert-Beer's law 68, 71
laminar flow 9, 17, 27 ff, 80, 154, 215
lamination 30 ff, 176
lapping 119
LASER 66, 71, 72, 225
– ablation 178
– Argon-ion laser 153
– bonding 177
– $CO_2$-laser 168, 201
– etching 153
– micromachining 173 ff, 206
– settings 163
– welding 176
LED 67, 73
levels of packaging 185
lift-off 133, 139, 140, 162, 197
Light Emitting Diode *see* LED
light source 66, 68 f
linewidth 123, 124
liquid crystal thermometry 64
liquid glass 166
loading 149
low pressure chemical vapor deposition (LPCVC) 132
luminescence 66
luminol 220, 222